Health and Healing in Rural Greece

Health and Healing in Rural Greece

A Study of Three Communities

Richard and Eva Blum

Assisted by
Anna Amera and Sophie Kallifatidou

1965
Stanford University Press
Stanford, California

Printed and bound by Stanford University Press
Stanford, California, U.S.A.
L.C. 65-13108

To Nevitt Sanford

Acknowledgments

We are greatly indebted to the Greek Ministry of Health and to the Attica Health Center for their cooperation and kindness in providing the public-health personnel and facilities employed in this study. We are particularly grateful to Dr. Elias Mavrolidis, Director General of Health for Greece, whose understanding support contributed so greatly to the success of our study. We are also indebted to Dr. E. Andreadis, Chief of the Cancer Control Section of the Ministry of Health, whose encouragement and good advice were invaluable; to Dr. Thomas Katsakos, Medical Director of the Attica Health Center; to Dr. Meropi Violaki, Deputy Director; to Drs. Evgenidis and Kritharis, Mmes. Charalambidou-Leontidou and Tavoutsoglou-Tolakopoulou, and Messrs. Sirmopoulos and Kourebanas of the Health Center staff; and to Drs. Stavrakakis and Karaklis and Miss Zografou of the staff of the St. Sophia Children's Hospital of Athens.

We owe much to the friendship and wisdom of Miss Litsa Alexandraki, Deputy Director for Liaison of the European Inter-Governmental Commission on Migration (formerly of the Ministry of Health and Welfare), and to Dr. Spyros Doxiadis, Medical Director of the St. Sophia Children's Hospital. We also wish to thank Mr. Michael Lustgarten, Mr. Nicolas Damigo, and Mr. Daryl Dayton of the United States Embassy, Mrs. Maro Holeva of the United States Information Service, and Miss Mary Stavropoulou.

Mention must be made of the extensive assistance we have received over the years from Professor Dimitri Loucatos of the Folklore Department of the Academy of Athens, and of the help we received in our recent field work from Dr. Vasilios G. Valaoras, Professor of Hygiene and Epidemiology at the University of Athens School of Medicine. Finally, we would like to thank the Nomarch of Attica, Leonidas Papa-rigopoulos, and the president, police-director, teachers, and priests of Doxario, who were delightful hosts and wise advisors.

In order to safeguard the identity of those who gave so generously of their time and of themselves, both the names and the places mentioned in this study have been changed.

Contents

Health and Healing in Rural Greece

Introduction

The aim of this book is to provide the reader with an understanding of the health beliefs and practices of peasants and shepherds in rural Greece. We believe that the understanding of any one set of human endeavors or activities must rest upon an appreciation of the larger context of life and belief in which that activity occurs. Consequently, our effort has been not just to give the facts on health and healing, but to emphasize and to try to integrate historical, cultural, social, economic, and psychological observations as these bear on what rural Greeks do to maintain life and strength and to ward off pain and death, for those actions are the essence of health behavior.

That emphasis is, we believe, a two-way street. By knowing how people live, what shapes their behavior, what they fear and desire, one can understand their particular health behavior. Conversely, by focusing on health and healing, one can learn a good deal that might otherwise be overlooked about the culture itself, and about the structure of communities, the nature of social relationships, and psychological propensities. Our aim has been fully as much to learn more about Greek culture as to learn more about health.

Our study began in 1957 with an inquiry into the historical antecedents to modern peasant beliefs. These inquiries continued during successive visits to Greece, and in five months of field work in 1962. The findings on health and healing are reported here; the findings of our analysis of the folklore of life and death will be presented in a subsequent volume. The present volume should be of interest to those concerned with health behavior, with ancient or modern Greek culture, and with peasant societies, as well as to behavioral scientists, folklorists, and those involved with the practical problems of community development, administration, and economics in underdeveloped countries.

OUTLINE OF THE STUDY

Chapter 1 gives the setting for the study, describing the situation, population, institutions, and major activities in the two peasant villages, Dhadhi and Panorio, and the encampment of Saracatzani shepherds, which were the communities we studied.

Chapter 2 takes a historical approach to the description of the Greek style of life. It is based on the recognition that rural life styles are not fashioned anew each generation, but rather that the generations follow the tradition-rooted schemes dictated by the natural environment, history, culture, and structure of the psyche. What was past not only has shaped the present but may *be* present; a full understanding of what is requires knowledge of what was. Thus, in Chapter 2, rural life in the time of Hesiod and Homer is described, and—so that one may readily see how little separates past from present—brief reference is made to a modern study of the Achaean-like Dinaric warriors. Out of that study of the past there arise several themes and phenomena of life style which are, to our way of thinking, fundamental to the Greek orientation, and which provide valuable insights into modern Greek culture and character. And, again, there is a two-way street; for as we propose that enlightenment about the present proceeds from the past, so archaeologists propose that knowledge of the present illumines antiquity. As the archaeologist William McDonald has observed in a recent article, "In more isolated areas of Greece, many aspects of life still reflect a pattern not too different from that of little agricultural communities of the ancient past."

Chapter 3 brings us up to date, not only describing themes in contemporary community living but emphasizing the revolution in life styles that has recently begun, a revolution that is likely to produce as much change in thirty years as has been wrought in the preceding three thousand. We describe the dominant patterns of social interaction, role behavior, and psychological response, and set forth some of the constants in man's relation not only to nature but to other men, constants that later will be found to color what goes on between patient and healer, and between the healthy and the sick.

Findings on health per se are presented in Chapter 4. Data on illness, as reported by families and compared with the findings of physicians, are presented, problems of taking medical histories and of underreporting in household surveys are discussed, and folk disease concepts are considered.

Chapter 5 presents information on birth, death, and abortion, and necessarily, in view of our initial aims and assumptions, leads to a consideration of infanticide, barriers to early marriage, birth-control methods, and beliefs about the cause of infant ills. The notions of the supernatural power of the infant and of his vulnerability to the evil eye also come under scrutiny.

Treatment activities are the focus of attention for Chapter 6. Healing efforts, folk cures, and problems of hospital utilization are considered. In Chapter 7 we begin an analysis of the action-research program, which introduced a team of public-health physicians, nurses, and laboratory technicians into each of the three communities we studied. How the villagers felt about the medical examinations they underwent and how well they understood and cooperated with the medical advice they received are the focal questions dealt with in this chapter. Chapter 8 presents the results of an inquiry into the villagers' hygienic knowledge and techniques, and reveals more about the beliefs upon which hygienic activities are based. Chapter 9 extends the inquiry into the causes and cures of illness, and what can be done to avoid it.

In Chapter 10, descriptions of and interviews with physicians, pharmacists, and midwives in Doxario are offered to give insights into the problems of providing medical care for the villagers. Conversely, interviews with the villagers provide information on physician choice, conceptions of qualification, and beliefs about the obligations incumbent in the doctor-patient relationship. Chapter 11 examines the pervasive patterns of use and belief associated with folk healing. The various kinds of folkhealers, their specialties, their work, and the nature and transmission of their power and skill are discussed.

Chapter 12 focuses on the person of the folkhealer. Using a case-history method, we attend to the person, the work, and the relationships with others of Maria, the wise woman; of Kostas, the hand-practikos specialist; of Mantheos the magician; of Dionysios the sorcerer; and of Vlachos, "the god." The priests are also folkhealers, and in Chapter 13 their position and healing work are described. Again the case-history method is used, and we introduce three very different men: Father

Dimitri, Father Manolios, and the one called "Father Terror." Their views of illness cause and cure are presented, as are their evaluations of their parishioners.

Chapter 14 concludes the book with a discussion of the broad national factors that affect the provision and utilization of medical care in the three communities we studied. Examples are given of problems in medical education, in the organization of health services, and in individual and bureaucratic responsibility. Difficulties that arise when Greek or foreign agencies engage in direct efforts to introduce health programs into rural areas are also discussed. Our emphasis continues to be on the need to understand the network of beliefs and social relationships within which one is working. Following Chapter 14 are an Appendix giving methodological details and two bibliographies, the first giving references cited in the text and Appendix and the second giving general references of interest to those concerned either with health behavior or with modern Greece.

METHODOLOGY

What is presented here is a general description of our methodology; methodological details are given in the Appendix. Without attempting to describe all of the methodological hazards to be encountered in the frontier area of health behavior and healing lore, we may usefully define certain of the limits of our own techniques and warn the reader of possible sources of error and misinterpretation.

We may appropriately begin with the problem of emphasis. In his brilliant book *The Little Community,* Robert Redfield considers the ways in which a social scientist may seek to approach and understand the life of people in little communities—those camps, villages, and small towns that are distinct, relatively homogeneous and self-sufficient, and in which over one-half of mankind lives. A major problem is to decide how one can describe a small community as a whole and at the same time isolate its component parts. A second problem is that any approach imposes a structure on the material being studied. Any imposition, whether it be a physical intervention in order to observe, or a logical category constructed in order to codify data obtained, limits as well as sharpens our vision. It highlights a feature, or a system, or a theme, and leaves others in shadow.

Most anthropologists and sociologists agree that peasant communities the world over are much alike. G. K. Foster, in *Traditional Cultures,* comments on their remarkable similarity. Nevertheless, various investigators have demonstrated a number of real differences and, more than that, have disagreed about what the important characteristics of these communities are. Redfield, in *The Little Community,* observed how much at variance he and another anthropologist, Oscar Lewis, were in describing the same Mexican village of Tepoztlán. Redfield saw a "relatively homogeneous, isolated, smoothly functioning and well-integrated society made up of contented and well-adjusted people." In his book *Life in a Mexican Village, Tepoztlán Restudied,* however, Lewis emphasized "the underlying individualism of Tepoztecan institutions and character, the lack of cooperation, the tensions between villages within the municipio, the schisms within the village, and the pervading quality of fear, envy, and distrust." Redfield thinks these differences in interpreting the same data arose from a difference in the underlying questions that the two investigators wished to answer. The hidden question behind his own study, said Redfield, must have been "What do these people enjoy?" Professor Lewis, on the other hand, must have asked "What do these people suffer from?"

Our orientation has likewise shaped our conclusions. Our major questions were "What have these people been like in the past, what are they like now, and how does what they do to stay healthy make sense in terms of their history and style of life?" It was our expectations rather than our questions that may have been "hidden" in Redfield's sense. What we expected is that healing is in fact integrated with the rest of the social order, and that both the social order and healing lore are shaped by historical as well as present environmental forces. These expectations and the methods we used to pursue them are derived from our own backgrounds. One of the writers is a clinical psychologist with training in classics and work experience in anthropology. The other is a social psychologist with additional interests in anthropology, sociology, and archaeology. Our two Greek associates were social workers, one psychiatrically and the other sociologically oriented. None of us are physicians, but all four of us have worked in public health, and two of us have worked intensively in the study of medical care.

The most difficult aspect of presenting results, that is, in describing the people of the village and what they think and do, is not in giving facts—reporting distributions of behavior or the presence of conditions that are well defined and measured—but in making abstractions from the observed data that present a picture of the whole. Abstractions integrate and encompass details, but they also distort and idealize; they increase understanding at the cost of limiting the information presented. We recognize that the chance for error increases as interpretations are made and overviews or themes or theories constructed. Nevertheless, we deem the gain worth the risk, and have proceeded, especially in elucidating historical and contemporary themes, to make inferences and interpretations and to offer theoretical formulations.

Each investigation of nature, including the nature of man and his societies, relies on instruments. In our study the most important instrument was the human observer. The four of us, our eyes and ears, our ideas, our ability to intervene in village life in certain planned ways and to record responses to what we did, were the tools. In addition, we obtained the help of physicians, laboratory technicians, nurses, scribes, and other public-health personnel. What they did, what they heard and saw, led them to make observations that also became our data, just as our observations on the interaction between these medical people and the villagers during the action-research provision of medical care generated additional information.

We used several standardized forms and procedures. One was an observation schedule that contained 140 topics or guidelines for inquiry. In addition, we used three interview schedules or questionnaires, one with 22 questions for medical healers, one with 19 questions for nonmedical healers, and one with 87 questions plus some subsidiary inquiries for each family in each community. The family interview schedule also incorporated the Brief Form of the Hygiene Scale for Families developed originally by Stuart Dodd for use in Syrian villages. We also employed a daily observation log, and for the medical examinations our physicians were supplied with conventional forms for recording diagnostic findings. Census reports and other governmental documents provided us with supplemental data.

There is much to recommend the use of a standardized questionnaire in community studies. It provides detailed information leading to an accurate composite picture of a village and, by using the composite as a point of reference, it shows where a given person or family stands in relation to his neighbors. It prevents the sampling distortions that occur when one relies solely on informants, who by their nature are exceptional people. Its disadvantages are several: it restricts spontaneity, it limits focus, and it is subject to a number of error sources—interview and respondent bias, memory failures, conscious evasion, and so forth. A more detailed discussion of the problem of questionnaires in household morbidity surveys is given in the Appendix.

The questionnaire was so long—taking over eight hours to complete in some cases—that we soon decided to ask core questions of everyone, and to split the remaining questions into two units, one unit to be asked by the interviewer of half the families of each village, the other to be asked by the interviewer of the other half of the families in the village. To avoid confusion about who was to see whom, we sampled families not randomly but geographically; one interviewer worked from one end or side of the village toward the center, the other from the other side. Among the Saracatzani the two physically separate camps were divided between the interviewers. Some bias was introduced by this sampling procedure; in Dhadhi, for example, a cluster of four families from Attica lives at one end of the village separate from the predominant refugee population.

By the use of such sampling of half of the total population, we arrive at figures that are estimates only. The accuracy of these estimates is a function of the sample size; the larger the sample, the less the error, presuming the lack of bias in drawing the sample. The error in sampling is also related to the variability of what is being measured. The likelihood that the sample replies differ from the replies that would have been given by the village as a whole decreases the more the replies are homogeneous or in agreement.

Of course, observations of any sort that are made within a village are subject to sampling error, although this is not generally a concern of those doing community studies. For the most part, the assumption is made that exposure over a long period of time provides enough opportunities for observation to permit the observer to gain reliable impressions. Raoul Naroll's article "Controlling Data Quality," in dealing with the reliability of statements in ethnographic field reports, suggests that the accuracy of field reporting increases with length of stay in the area, increases when the observer knows the native language, is greater for scholarly observers, is improved by participant-observer techniques, and can be accepted with greater confidence if reports are consistent or independently corroborated.

We four observers lived in the field for a total of nearly 18 man-months. Three of us spoke Greek, two of us were native professionals, all of us were participant observers; and since our reports were both consistent with one another and corroborated by other studies, it is likely that our interview samples approached adequacy. Naroll claims that denial by informants is one of the greatest error sources, a point that he illustrates by the finding that denial of witchcraft practices often leads to failure to report this phenomenon. Our data—both interview and observational—show no scarcity of reports of the witchcraft phenomenon, a fact that encourages us to be confident of our results.

A special difficulty in attaining reliability arose not so much from the methods employed but from the changing nature of the communities themselves. Unreliability is introduced not only by the very nature of the questionnaire or the observation methods, but also by the instability of what is being observed. For example, it was never possible during the observation period of this study to arrive at an exact population figure for any of the three communities. Babies were born, sick people entered and left hospitals, relatives moved in and out, workers migrated to fields afar and returned from journeys, and so forth. Methods attuned to gaining static impressions produce data that are immediately out of date; on the other hand, methods attuned to describing change fail to emphasize the stable states. We accommodated as best we could to the dilemma by acknowledging our static figures as estimates and suggesting that most short-term changes do not alter basic social structures.

The clinical medical examination presents essentially the same advantages and disadvantages as the standardized questionnaire. It evaluates each person according to the same criteria and by means of the same instruments, and it does so systematically. Accordingly, one learns much about a person and about how he compares with others, and, after accumulating the data, one can describe the distribution of health conditions throughout the village. But the medical examination is also subject to errors of both reliability and validity. Certainly there were circumstances that may well have produced error in the medical examinations as they were conducted in each community.

For one thing, the investigators did not have funds available to hire doctors to conduct the medical examinations. The Ministry of Health, under directions from the Director General of Health, Dr. Elias Mavrolidis, was generous enough to give its support to the study, and to provide two public-health teams, each consisting of a physician, a nurse, and a scribe, so that a morbidity study could be conducted. There was an eleven-day field assignment for the two teams. The working day was from four to six hours, which meant that approximately 300 people (for some people from surrounding villages presented themselves for free examination, thereby adding to our villager sample of 262 persons), or 150 patients per doctor, were seen over eleven days during which the working period averaged no more than five hours a day, or just about twenty minutes per patient. That this amount of time would be considered sufficient for taking the case history and making the clinical examination of a new patient by American standards seems unlikely.

Another problem in our morbidity survey was the limitation in equipment. Neither an ophthalmoscope for examining eye grounds nor an otoscope for examining the ears was available to our team. For a further discussion of methodological problems in the medical examination, see Appendix I.

To summarize our methods we used four major techniques and a number of subsidiary instruments. One technique was that of the participant observer. It was based upon our living and working in the region of study. The second technique was the standard stimulus situation—in this case, a formal interview with each household in each community under study and with each healer likely to be visited by members of those communities. A sampling modification was introduced in that one extension of the questionnaire was asked of half the families in each community, another extension of the other half. The third technique was the spontaneous interview, gently guided, whenever or wherever the investigators and the villagers or healers were together. The fourth technique was the combination of action research and a natural experiment. We organized the communities for participation in medical care, introduced public-health physicians, observed the physicians and villagers together, and followed up on the response of villagers to medical advice. This allowed us to work with both private and public agencies as we prepared for the morbidity survey, and to observe those organizations at work. Further opportunities for observation arose as people from our communities traveled for examination and treatment to Athenian clinics, and to distant folkhealers—priests, nuns, magicians, and practikoi. Through these methods we also became acquainted with local physicians, pharmacists, and midwives, and with people in the capital, government officials, city physicians, university faculty, and others in the urban and the rural elite.

1

The Setting

The two peasant communities and the shepherd encampment that we studied are located within the same region and are under the jurisdiction of a central community, which we shall call Doxario. Each community differs from the others in important ways, primarily in ethnic background and economic way of life.

The largest community, Dhadhi, has a population of about 200 year-round residents living in 50 family households, and is located in a fertile plain. Most of its people have their roots in Asia Minor, for they or their parents came to mainland Greece in 1922 after the disastrous war with Turkey, to be resettled in what was then a great swamp. Many still remember Turkish and use a few Turkish words. Most of these people are peasants; they make their living by raising wheat, grapes, and garden produce, and by selling milk from their cows, as well as by earning wages. Each family that settled in Dhadhi in 1922 was given ten acres of land, plus grazing rights on about ten acres jointly held in community lands. Changes have taken place since this original distribution, so that now some families own more land, whereas others have sold theirs or passed it on to their children through inheritance and dowry settlements in fragmented parcels. A few families, newcomers, own no land at all. The largest holding for any one resident in Dhadhi is about 20 acres, although several absentee landlords own much larger holdings on the periphery of the village.

The average family lives in a two- or three-room flat-roofed, adobe-walled house. None of the houses has electricity; only the teacher's house has an indoor toilet. The typical family has its own water well within a few yards of the kitchen. But the well is usually located only a few feet from the privy, the only waste-disposal method in use, and, if the family owns cows, near the small stable adjoining the house. Many villagers have small propane stoves for cooking, but only the tavern keepers have ice boxes. The families own a number of items—glassware, furniture, knick-knacks—made outside the village, and all of their clothes are bought in the city.

The total income of these people is difficult to estimate because the reports of villagers conflict. Probably little more than half of their income is in cash; the rest consists of home produce that is consumed but which can be estimated at market value. Our estimate, based on expert and local reports, is that an average family or household has a total income of slightly over $1,100 per year, of which half is cash. This conjectural figure is slightly higher than the average-per-family rural income for Greece, which is about $1,000, as reported by Papandreou (49). The range in income within the village is great, varying from about $300 a year to several thousand dollars.

The institutions of Dhadhi are the family, which is the central social unit; the school, whose teacher is the most respected man in the town; the taverna–coffee shops, where the men gather to drink and talk in the evening; the agricultural cooperatives, of which the milk cooperative, directed by a council of men in the region, is the most important; the village council, which is a relatively unimportant group, and the church, which has no regular priest and is rarely attended. Because Dhadhi is an outlying hamlet under the jurisdiction of Doxario, it has no local government officials, not even police. Its officials are those of the community of Doxario, a district or region covering perhaps 750 square miles and having a total population of 3,090 persons in 1961. This community has an elected president and council who are responsible for the local affairs of the six villages within the district. It also has a police station whose gendarmes are members of the national police force under assignment from the central government.

Two priests live in the central village of Doxario, seven miles by road from Dhadhi. Their salaries are paid from national tax monies for the purpose of ministering to the religious needs of all the villages in the district. In Doxario also are all the health personnel who serve the region—two doctors, two pharmacists, and two midwives. The district has no hospital or medical laboratory.

Because the village of Doxario is the commercial and administrative center for Dhadhi, the road connecting the two is a vital force in the lives of all of the villagers. The road is rock and dirt from Dhadhi to the intersection of the paved highway to Athens, about a mile outside of Doxario. It is traversed once a day by bus, which makes a quick round trip in the afternoon, and more often by donkeys, pedestrians, farm trucks, and bicycles. Most of the vehicles on their way to Doxario are owned by the milk cooperative. A few are owned by traveling vendors of household wares. One or two of the villagers own motorcycles, and one has an old truck. All have donkeys.

No one in Dhadhi except the teacher has genuine, formal authority. His high status—visible in his air of confident superiority, his good clothes and fine house, and his charming and educated wife—enables him to assume powers beyond those defined in his teaching position. He has taken the initiative in community action, pushing and pulling the sometimes reluctant villagers to make improvements in the school and the church, and campaigning successfully for the extension of electric power lines into the village. His approach is energetic, authoritarian, and arrogant; his effort is not to coordinate or organize the village, but to push it forward like earth before the blade of a bulldozer.

There are two men of means who are respected in Dhadhi. One is a resident of Doxario who keeps a large stable of cows in Dhadhi, tended by a hired nonrefugee family. The other is the landowner, a member of one of Greece's great families, upon whose expropriated and donated family holdings the village was built. Through government expropriations in the last 50 years these holdings have been reduced from many thousands of acres to 500, but for people whose plots average ten acres, 500 constitutes a domain. The owner of these lands is a great man in the eyes of the people, one whose role corresponds to that of the aristocrat or noble in the northern European countries. His very real benevolence has provided Dhadhi with a school and a church. And through his potential capacity to intervene with government authorities on the villagers' behalf, he gives them also a sense of protection.

Panorio is the second community in our study. It is about eleven miles from Dhadhi and three miles from Doxario. It can be reached by a paved but narrow and winding road that climbs into the mountains, its edge dropping off precipitously into ravines. Panorio is a small village with about 126 residents, most of whom are peasants but a few of whom are shepherds. There are 29 family households. They are ethnically homogeneous, stemming from two or three families or clans that came to Greece during one of the waves of Albanian migration. Nearly every family speaks some Arvanitiki, an Albanian dialect. The time of their migration is uncertain; they have no recollection, but one historian (8) thinks the most likely date is the fifteenth century. Despite their northern origin, they think of themselves as indigenous folk, and indeed some historians, including Biris (8), classify them as Dorians. Perhaps their continuous contact and marriage with indigenous people from nearby mountain villages has been responsible for their absorption of the local traditions that hark back to antiquity.

The villagers of Panorio recall the transition from the shepherd's life that took place in the nineteenth century; the few among them who still herd sheep or goats carry on a pastoral but not nomadic way of life. The women of the village weave clothes and linens which are very artistic, reflecting styles and patterns in use for centuries. The men and younger people wear city-made clothes. The majority of the villagers live in two-room houses, most of which have sloping roofs of wood and straw covered with tile and stone. Architecturally these houses appear to be halfway between the beehive-shaped huts of the shepherds and the houses of settled village folk.

They have no plumbing or electricity, and unlike the people of Dhadhi, they have no privies or family wells. Except for one family, they get their water from one of several community wells situated at the bottom of the hill on which the village is located. Their food is cooked over wood fires, and in winter their heat is provided by the single fireplace in each dwelling.

The average family owns fewer than ten relatively unproductive acres, but three have nearly 25 acres. Some of the younger people have none; until they come into their fragmented inheritance, they must work for someone else. The villagers grow wheat, grapes, beans, potatoes, peas, and olives. Nearly every family has a few chickens and goats.

As in Dhadhi, estimates of family incomes in Panorio are difficult to make, for experts and villagers give conflicting reports. The best estimate for total income, of which about half is probably cash and the rest home-produced and home-consumed food, fuel, and clothing, is about $900 a year, with a range of $300 to $1,500. This conjectural figure is slightly less than the Greek average for rural family income.

Like Dhadhi, Panorio has no local authorities and is under the jurisdiction of the central community of Doxario. Although there is a newly built school in Panorio, the teacher does not live in the village, preferring the more comfortable life in the village of Doxario down the mountain. Consequently, he plays no part in village life aside from his job within the walls of the one-room schoolhouse.

Within Panorio there are two families who are considered wealthy by village standards. The fathers of these families have both been the elected president of the local village council. Both are related to most of the families in the village, and consequently their positions were a result of kinship as well as demonstrated community interest and wealth. In return for their helpful activities they received a limited amount of respect and authority—limited because envy and competition, fierce individual independence and group factionalism, were so intense that very little real power accrued to either leader. Both acknowledged their inability to organize the village or to persuade others to cooperate effectively or pool family power and resources. Both spoke of their discomfort in the face of the resentment, jealousy, and verbal attacks their higher status brought them. It was apparent that envy, factionalism, and rebellion reduced community government almost to anarchy and prevented many measures for the common good.

In spite of these social and psychological barriers to community development, the village reflects the material progress of the surrounding region. Within the last decade a school has been built, one of the churches enlarged, and a paved road put through the village. The greatest local contribution of land (for the school and road) and supplies (for the church and school) came from these two families. We were told that many of the villagers had been opposed to the road because the route endangered their tiny holdings, and remained hostile to it even after one of the town leaders offered his own land for much of the right-of-way. When the road crews arrived the villagers stoned them, and, reportedly, only the intervention of the two leading elders prevented the road workers from being seriously harmed.

There are no cooperatives in Panorio, and its less productive land allows the villagers fewer commercial transactions with outsiders than the people of Dhadhi have. Buses pass by the town four times a day going toward Athens, an hour and a half away, and toward the mountain communities to the north. Local young folk use the buses for transportation to their work in the fields on the plains below, and occasionally a truck belonging to a cooperative comes into the village or one of the merchants from the outside sends one to pick up marketable produce.

The most important institution of the village is the family. Since nearly every villager is related to ten or fifteen others, one might expect the kinship ties among cousins to provide a network of solidarity. But this is the case only in the sense that one's cousin is not a stranger; he is not necessarily trusted, nor is he always a friend. Close ties exist primarily within the household unit, or between households when the parents live next door to the families of their children.

Panorio has two churches, one located on a winding path in the center of the clustered houses, the other, larger church standing on the far edge of town. There also, on a site sacred since antiquity, is the cemetery. The village has no priest, but once a month one of the priests who reside in Doxario conducts an ill-attended Sunday service.

There are two tiny coffee houses, or tavernas, in the village, both run by peasants as a sideline. Men gather there to talk and to read the paper or, if they are illiterate, to listen to others read it aloud. Neither shop is imposing, and only a few men gather there on occasional evenings. As in Dhadhi, the coffee houses in Panorio offer odds and ends of essential merchandise for sale: wine, canned fish or tomato paste, needles and thread, cigarettes, salt, and the like.

The third community, the shepherd encampment, is a dispersed one. The eight families, composed of 42 persons, live in thatched huts scattered on the mountainsides near Doxario, overlooking the plains and the sea. The people of this community are the Saracatzani, shepherds from northern and central Greece who migrated into this region only within the century, people who are descendants, it is said, of the ancient Greeks (31). These are people who carry with them in everyday life traditions of which Homer sang. Achilles' warriors would be at home among them.

The sheep are their life; their calendar is the shepherd's calendar, with the year divided into two periods—when the sheep are in pasture and when they are not. But within the last few years the Saracatzani have begun to settle. They are building houses in towns and abandoning their nomadic life to herd the sheep in restricted areas, where they are building the semi-permanent huts of their camps, raising a few cows and chickens, and even renting nearby land to raise a little wheat. They have sold their horses and abandoned their annual mass encampments, their county fairs, and their traditional assemblies during which marriages were made and leaders selected; they have stored their lovely costumes away and can be seen in the very midst of transition from nomadic to settled ways. To the older folk the changes are regrettable but necessary; to the younger people the changes are desirable and only too slow.

The Saracatzani encampments are family groups, everyone on the mountainside being at least a cousin to another. As a clan they live intimately and cooperatively, each sharing in the rugged shepherd life. The huts, of straw and wood, are shaped like beehives with a triangular opening in each just beneath its ridgepole for the smoke to get out. They are not unlike the stylized tombs of the Mycenaean-Minoan civilization, which are located on the plains below the encampment.

Of the three communities, the encampment of the Saracatzani is the most primitive. Hut floors are of mud; there are no chairs, beds are few, and only the simplest utensils and possessions are kept in the camps. Although the preparation of some of their special food dishes is unique, the staples are like those of the other people in the region: bread, vegetables in season, olive oil, goat cheese and milk, a few eggs, wine, and occasionally meat or fish.

It is likely that appearances indicate greater poverty than actually exists. According to an expert rating, a total family income from a flock of 100 sheep (a number slightly above average), a few cows and chickens, and the growing of a little wheat, plus the wages of the young people and the women working as seasonal laborers in others' fields, is about $1,000 per year, with a range of between $500 and $1,300.

The Saracatzani have few institutions other than the immediate family and the clan extensions. Within the family, leadership is exercised by the eldest male. In the past there were chieftains who led the nomadic groups and were responsible for negotiating land leases for sheep grazing. With the recent settling of the Saracatzani and their breakup into smaller family encampments, the chieftain role is disappearing. Residuals of the old forms can be seen in the sheep cooperative, which is a modern device to perpetuate the sheep-based economy, and in the selection of one elder brother as "secretary-treasurer" of the cooperative, in which role he collects money from the other families and negotiates the land leases, thus carrying on the chieftain activities. As for other institutions, the Saracatzani must leave the mountainside for the villages to find church, coffee house, school, or community council.

SOME MUTUAL COMPARISONS

In each of these communities the primary social unit is the family. The household often includes at least three generations, sometimes four. It may also contain unmarried or widowed sisters of any of the elders, as well as an adopted "soul-child"* or a more distant cousin or a nephew who has come to stay. In spite of the spanning of generations, the average household size in these communities is only four persons, with a range of one to eleven (a few bachelors live alone, but no woman does). The proportion of children in the household is highest among the Saracatzani, where over half of the average family are 16 years of age or under. There are the fewest children in Panorio, where only one quarter of the population is 16 or under.

The children in each community are enrolled in primary schools that provide six years of education. For the Saracatzani children, going to school means walking two to three miles, but they all make this trip every day. Although there is a secondary school in Doxario, many of the children will not go on. Still, they do get elementary schooling, whereas many of their parents have not been as fortunate. Among the Saracatzani the majority of the household heads have not gone to school at all, while in both Dhadhi and Panorio the majority have had less than four years of primary school. The adults are unhappy about this, for education is highly prized. Many adults do not know how to read, and their information about many aspects of life is limited. In regard to health, this limited education results in a lack of knowledge of body functions, sanitation, nutrition, and hygiene.

Of the three communities, Dhadhi is not only the largest but also the most heterogeneous. Whereas all of the Saracatzani are related to one another, and nearly all of the people in Panorio have kinship ties with other households, one-fifth of the people in Dhadhi are unrelated to another household, and one-fifth are related to only one other household by blood or marriage. Their heterogeneity is seen in their ethnic background, for in spite of their predominant origins in various towns in Asia Minor, there are newer settlers whose families are from Albania,

* Either a child or an adult living with a family, informally but publicly "adopted" as a family member.

the Greek islands, Attica, and the Peloponnese. Their diversity is also expressed in the geography of the village, for its 50 families are spread out along a single row of houses forming a loosely strung "T" that extends over several kilometers. At the farthest ends of the "T" the newcomers from Attica, both peasants and shepherds, are grouped. Just as the people of Dhadhi are not physically close, they are not socially or psychologically close. As a consequence, communication among them is practically nonexistent; they live independent lives, which, while not always cordial, rarely involve intense dislikes. Within such a community one can expect to find a wide range of views and practices, especially in matters unrelated to morality or orthodoxy, or in matters that do not stem from limited common experiences imposed by the restricted geographic and economic conditions.

Panorio is more homogeneous. Most of the houses are grouped together, and ties of kin and a more restricted life closely bind the villagers. The village is small; the trip to the community well demands social exchange; the route to the main road requires passing down the central path of the village. Communication is speedy, although not without distortions introduced by drama and prejudice. Relationships are intense, and have a hazardous potential for enmity. There is neither distance to mitigate anger and envy, nor acknowledged leadership to control their expression. Anger and envy are modulated by withdrawal, crystallized as people ally themselves into factions, or directly released through immediate verbal, and occasionally physical, aggression. But whatever their intensity, the people of Panorio are not savage; good humor and wisdom are their salvation. As a result, the fabric of their daily life is colorful, uncertain, and often distressing, but throughout it a sense of having to live together come-what-may prevails.

The group of Saracatzani that we studied comprises two encampments; although they are physically separated by two mountainside miles, theirs is a single community without social distance, bound by ties of blood, marriage, and way of life. One of the two camps has five households, the other three. Extra beehive structures for sheep, goats, and chickens are scattered about, and there are common ovens—domed structures of stone and cement—around which the women gather when they bake. The women sit outside when the weather is mild to spin, weave, or repair their hardy homespun cloth; they often walk together beside their donkeys as they labor up the steep slopes, carrying water in jars or milk cans from distant springs. They walk together when they take their husbands, the shepherds, their evening meal on slopes high on the mountains; they return in the darkness carrying milk from the sheep. Perhaps it is because the families are such close kin, or because the shepherd's life—with all its hardship—provides a safety valve in the solitary tranquility, that the Saracatzani community is tightly knit and harmonious.

2

Themes in Community Life: Parallels, Past and Present

Health beliefs and practices must be viewed within the context in which they occur, since focusing on them in isolation distorts or detracts from their meaning and function. Therefore we shall present in this and the following chapter a brief sketch of the major life themes and personality characteristics that shape and color the health beliefs and behavior of Greek peasants and shepherds in the communities we studied.* Our discussion is intended to provide a perspective from which may be viewed the total fabric of life in which attitudes and actions are imbedded.

We shall begin with an interpretation of Greek village life by the Greek Hesiod (23, 32), who wrote some 700 years before Christ. In Hesiod one finds value placed on prudence and piety, on work, on vigor, on minding one's own business, on moderation, on practicality, on respect for the elders, on constrained individuality and modulated competition, on submission to the rulers, on righteousness and tradition, on thrift and planning, on the accumulation of material goods without recourse to commerce, on truthfulness, on neighborliness and respect for one's community obligations, on well-earned leisure, and on reserve in speech and restraint in social relations. Hesiod, who in *Works and Days* describes the ideal rural life, is critical of extremes in behavior—idle-

* For a more complete observation than this sketch provides—for breadth, scope, and a comparison with our findings—we recommend the following excellent works on Greek culture: Leland Allbaugh, *Crete: A Case Study of an Underdeveloped Area* (Princeton, N.J., 1953); Ernestine Friedl, *Vasilika: A Village in Modern Greece* (New York, 1962); William McNeill, *Greece: American Aid in Action 1947–1956* (New York, 1957); Irwin Sanders, *Rainbow in the Rock: The People of Rural Greece* (Cambridge, Mass., 1962); and John Campbell, *Honour, Family and Patronage* (Oxford, 1964).

ness, greed, envy, poverty, violence, adultery, perversions, oppression by the rulers, gossip, litigation, bribery, sensuality, and falsehood. He contrasts these faults with a description of the ideal community, in which man shows piety toward the gods, rendering them sacrifices and avoiding offenses to them and leading a life in keeping with the natural order, with the gods' support. In such a community men cooperate but avoid becoming involved in one another's affairs; the wise man keeps his own counsel, but reserves his best behavior for his family and treats them well. Even in the good community a man is right to be especially distrustful of women, for they may be deceitful and polluted, endangering one's prowess or seeking one's property. Into this community may come strife, taking the harsh form of war and slaughter, or of mischief, which serves no good. But strife may also occur in the form of controlled envy, which leads to competition, emulation, and thereby to work and gain; this form, says Hesiod, is a force for the good.

Life is troubled, writes Hesiod, but work, foresight, and propitiation of and respect for the gods will do much to insure food to ward off famine, and protection to ward off illness. Work is hard and wearisome, pain and anxiety are common, but the unity of the family, the fruits of the land, and pride in good conduct and in material possessions provide great pleasure. With one's fellows one must be very careful: one must not be so foolish as to expose oneself to ridicule or shame. One must abide by one's duties of hospitality and mutual obligation within the village. There are dangers that loom from outside the community; falseness on the part of the feudal barons, or plundering warfare engaged in by violent men who recognize no right except that of strength. One can only submit to such people, but perhaps foresight would have led one to put something safely aside, or, if one has been pious toward the gods, one might expect their potent aid against marauders. The gods themselves may decide to punish the irreligious with famine or sickness or infertility; if one has offended the awesome ones who inhabit the night or the wild places, one may well expect to be struck down. Failure to observe the rites of purification, the sensitivities of the spirits, a disregard for the taboos that protect against pollution, are all dangerous omissions—omissions that will bring disaster to the offender. All things done and left undone are observed: "The immortals are close to us, they mingle with men," and omens of their presence are to be seen. Wise men guide their actions by this knowledge. They will not transgress against the deities or the powers of nature; they will live in peace with their fellows, abiding by the traditions that keep the community whole and displaying those personal virtues admired by both men and gods.

Hesiod's older contemporary Homer, who wrote of the warrior's life in an earlier Achaean-Mycenaean age, showed that the conflict and violence that Hesiod criticized were life-themes. Conquest and seizure were means for gain, and migration and adventure were paths to glory. Homer's heroes lived in tense hierarchies, with quarrelsome kings and assemblies of the passionate-tempered nobles. A man valued himself in terms of the opinion of others; his status was bolstered by fine speech, versatility, and cleverness, but ultimately it depended on his high birth and the swiftness of his sword.

The warrior had but few honorable alternatives. He had little chance to master circumstances. The poignancy of an encounter with his own vulnerability was sharp. Men were precipitated into internecine warfare by forces outside their control; their only choice was to be cowardly or to be brave. No wonder they vacillated; no wonder they boasted of their prowess while avoiding battle whenever they could do so without forfeiting the esteem of their comrades. In a situation where one's manhood, pride, and very life were always in question, it was always convenient to throw the blame on others, using gods and comrades as alibis. Fear of being shamed, of being found unmanly, of being a laughingstock —these were the forces compelling heroic and rebellious deeds. A man who was resigned to die, like Odysseus, might develop alternate virtues: good counsel, cunning, and patience; but these gained him respect only as long as his courage in battle remained unquestioned.

Is there an unbridgeable gap between Homer, whose song celebrated heroic times long past, and Hesiod, who moralized about the Georgic way of his own age? Perhaps not. On examining the *Iliad*, we see parallels to Hesiod. We find a king guided by a council of chiefs (12), deliberating before the periodic gathering of the assembly of the people. Just as in Hesiod, the family is the basic unit, and family ties bind without regard for affection—witness Menelaus and Agamemnon, Paris and Hector. Families live together in villages of their clan (genos). They are part of a larger tribe or kingdom (phyle), building a society on a pattern probably generated centuries earlier in northern and Asian heartlands from whence a hundred migratory streams had once flowed and would again flow into Greece.

Parallel to the kingdoms and passions of men in those days of the battle for Troy was the world of the gods, a bright mirror upon which human events were projected; the gods multiple, inconsistent, changing, comprising a religious structure that reflected the society of men in which those gods daily intervened.

Gods and traditions guided men, and king and priest and assembly interpreted the sanctions that the gods might wish to apply. Bloodshed and affront were family affairs; justice was arbitration of conflict in the light of the exigencies of power and of the sacred order binding man to god and nature. Conflict arose when one man threatened another's self-esteem or his family honor, or when robbers, be they kings, pirates, or migrating peoples, sought to seize the wealth—the land, the flocks, the fruit of the soil, the cities—of others.

Men visited hurt upon one another, but the gods visited ills—sometimes because of man's clear violation of the sacred order, sometimes because of the gods' whimsical cruelty. One might hope to fend off illness by ordering one's life according to traditions sacred to the gods: one could avoid offense to streams or render the sacrifice demanded; one could pray; one could read the omens—the flight of birds, for example. One could engage in magic, although neither Homer nor Hesiod shows much interest in its practice. Or one could—as one must—resign oneself to the unknown and uncontrollable, holding oneself ready to seize the opportunity when it appeared.

Homer's warriors and Hesiod's peasants had cultural elements in common despite differences in time, in migratory ethnic stock, and in predominant activities. Throughout the history of Greece there has been an intermingling of cultural groups. Whether through peaceful migration or through conquest, the newcomers settle and marry the inhabitants; the rural folk mix with those in the towns and cities; and the new gods are established—not without conflict—next to the old. The result is a larger fabric into which each group is woven. Some separateness is maintained, depending on a group's own heritage, environment, and style of life; but diffusion and accommodation blur many of the earlier distinctions, and the culture grows more complex, allowing new generations a greater choice of elements upon which to build and to which to respond.

These potentials, whether cultural or personal, inevitably provide for diversity and conflict as well as for sharing and homogeneity. Insofar as the way of life remained unchanged and the environment unaltered, no technological revolutions occurred, and migrant streams flowed only from old Illyrian and mid-Asian sources, one might expect that the elements present in Homeric and Boeotian epics continue almost to the present day—not just in Greece, but in the surrounding areas. It is our belief that this has been the case, and that except for two major changes—one, the introduction of the Greek Orthodox religion, in which the saints replace the old gods; the other, the recent industrial and now technological revolution that places increased emphasis on commerce and money, and arms the peasant with concepts and tools that will one day work a social revolution—the people of the villages that we have studied show substantial similarities to the peasants of Hesiod's time, and also reflect personal qualities seen in Homer's warriors.

THE DINARIC WARRIOR

The Dinaric shepherds are a contemporary people who inhabit the mountains that stretch through Yugoslavia almost from Switzerland to northern Greece. Their homeland, which was once called Illyria, is said by some (8, 12) to be the homeland of the Dorians, who were closely related to the native Pelasgians and to the invaders of Greece in the third millennium B.C. The Dinaric shepherds are related not only to the Dorians, but to the Arvanites of Panorio, and possibly through ancient Pelasgian ties to the Saracatzani.

Tomasic has studied the Dinaric herdsmen (67). Among them the family is the basic unit. The father is king; he is seen by the children as a god. The children are alternately spoiled and commanded; the results of such upbringing are submission mixed with defiance and intense love mixed with intense hate, emotions that spill over occasionally into infanticide and patricide. The Dinaric shepherds are a volatile people who move from resignation and passivity to violence, from excessive hope to dramatic despair. They are fickle, fast friends one moment and traitors to their companions the next. They are martyrs and tyrants, heroes and cowards. Physical force and tradition are their laws: "He who has a knife has a life," they say. Their gods are like themselves: God is "that old slayer," and neither a god nor a man is trusted. Blood revenge is demanded for bloodshed; not to exact vengeance is to lose all honor; to exact it is to condemn oneself and one's kin to ceaseless fear. They boast of their prowess before fighting, but they much prefer invective to blows, and given the opportunity will turn tail and run in battle. Cunning and deceit are common maneuvers; theft is the acknowledged means of gain; but he who catches the thief may kill him with impunity. To be a bandit is another thing—the bandit is respected. Magic is another means of gain, and women excel in sorcery. The people are superstitious and believe in omens and signs.

Women are subjugated but protected; girl children are without value. Marriage functions to beget and rear children, to maintain the economic and social functions of the family, and to cement the relationship between the tribes. Any affectionate attachment that may develop is incidental. Money is a means to status, and wealth is ostentatiously displayed. Life itself is composed of alternate moods and periods; one seeks peace in order to survive; one must face conflict in order to get ahead.

POLARITY

A summary of similarities between past and present, between shepherd, peasant, and warrior, has been presented. We now turn to some of the more specific life themes and personality characteristics that emerge. A pervasive feature of Greek life, past and present, is its polarity, the oscillation between two extremes—excess and moderation, chaos and order, anarchy and loyalty, brutality and tenderness, cooperation and competition, reverence and exploitation, distrust and sociability, overweening pride and fearful denial of good fortune, despair and fantastic hope, indifference and passion, cowardice and love of honor, female worthlessness and male godliness, family loyalty and disruptive selfishness, conservatism and openness to innovation, corruption and integrity, and cunning and frankness.

This theme of contrasts is one of the most impressive aspects of village life in Greece, one that appears to have existed throughout Greek history. It runs, as we have seen, through Homer's accounts of the ancient warrior's life, through Hesiod's description of rural life in ancient Greece, and through accounts of the Dinaric shepherds. It is this polarity that may have inspired the Greek dramatists, infused the character of Achilles, and pitted Hesiod against his brother Perses, and which today often pits the peasant in a struggle against nature, the community, and himself. As a fundamental characteristic of the Greek personality and expression, this polarity affects many aspects of the contemporary Greek villager's life, among them his health beliefs and practices, which are full of inconsistencies, contradictions, and even paradoxes.

EXCESS AND MODERATION

Polarity implies the capacity of persons to act in contradictory ways, believing in things and acting in ways that are opposite in their logic, assumptions, style, and effects. Polarity also suggests a potential for extreme behavior—behavior that is characterized by freedom from restraint, by intensity, and by a purity of theme. The villagers we studied exhibit such polarity, but their life is by no means extreme in all respects, for they advocate the virtues of moderation. Yet it is the initial contradiction of the villagers' lives that although they emphasize the ideal of moderation, and to a great extent act in moderate ways, they are often guilty of excesses and extremes.

The warning "Nothing to excess," from Solon and the Temple of Apollo at Delphi, is as necessary for the contemporary Greeks as it was in ancient times. This is not because moderation, the Apollonian way (7, 46), is so satisfying in itself, but because the Dionysian rapture is so appealing and yet so inappropriate to the requirements of the peasant's daily life. The peasant, with eight or nine thousand years of settled living behind him, has learned that farming and ecstasies, child-rearing and bloody battle, sheepherding and inebriation, do not mix. He may prefer the latter in each case, but he must live by the former. Therefore he seeks institutions, sanctions, and personal values that will remind him of the golden mean and protect him from his own—and others'—excesses. His proclamation of moderation is a bridle to govern his potentially dramatic actions. The outcome is a compromise: a life according to moderate values that is relatively stable when the sanctions against excess are clear and present and when moderate ways are demonstrably successful in satisfying the ordinary wants of food and affiliation, esteem and activity. The compromise becomes unstable in times of stress or boredom, when frustration or passion appears, when the sanctions of fear or solidarity dissipate, or when personal joy, challenge, or fulfillment dispel dedication to moderation.

As observers of the degrees of moderation and excess among the villagers, we must remember that their definitions may not be the same as ours. A man may say he has eaten only a moderate amount of bread, by which he means two or three pounds at one meal, or that he drinks a moderate amount of water each day, by which he means a gallon. Conversely, his hypochondriacal concern with a case of the flu may lead him to complain of extreme pain, whereas his newly fractured arm, which hurts but does not frighten him, "pains only a little." But we do not refer to these idiosyncratic definitions when we speak of moderation and excess. We refer to the contradiction of the village sentiment that one should never drink more than a few glasses of wine because to do so is dangerous to health and order, and the admission that people do drink too much, the results of which can be seen in the form of cirrhosis, the blind staggers, and the occurrence of knifings.

We refer to the sentiment that all foods are good in moderation, which contrasts with the fact that people gorge themselves and overfeed their infants; the idea that thinking too much causes madness, in contrast to the prevalence of preoccupation and rumination over family and subsistence problems; and the sentiment that peaceful neighborliness is the best way of life, in contrast to the intermittent eruptions of feuding and violence expressed by the village description, "Here we eat one another!" The consensus of belief is that one maintains health by avoiding excess, by following the rules of moderation in work, emotion, thought, dress, hygiene, and eating, and by observing the decrees of common sense, ritual, and taboo. Yet when villagers fall ill, they commonly blame their illness on their own excess in eating, drinking, working, worrying, suffering, and fighting, or on the flagrant violation of their own rules for the observance of rituals, taboos, and sensible precautions against the hazards of life. These explanations are not, as one might expect, self-recriminations over a moral failure to follow a life of moderation. They are merely explanations of the discrepancy between the ideal and the reality.

CHAOS VS. ORDER

The Greek myths reveal conflicts between polar antagonists—champions of creation, order, activity, and life battling the underworld, the primitive, unelaborated powers of chaos and disorder, and death itself. Yet as Fontenrose makes clear in his study of combat myths (21), the myths are not black and white. Hero and villain are not so easily distinguished as one might think; individual attributes shift, and opponents may even exchange their roles. "All gods are alike," says Fontenrose; "they can work both good and harm." The forces of order and of chaos, in part, lie within each antagonist. Whether the god is good or evil depends on what he is doing. The gods may change because their loyalties change, or because they are whimsical. But insofar as they themselves do not succumb to death, their power remains and is neutral. Its effects are good or evil depending upon the judgment of the viewer.

What does this mean for the villager? The names of the mythological antagonists have changed, but their effects are the same. Apollo's successor is St. George and Python is now the dragon, but the action is the same. St. George can bring healing, but when he is vengeful he can bring disease and disaster. The dragon, chthonic survival from ancient times, represents destruction and evil, and, in another form, is the snake of death, which, according to the Orthodox church, Christ stepped upon to kill, thereby securing everlasting life for men. But that same serpent (or his brothers) also plays a protective role: he is the house snake in the homes of Doxario, and his passing near is regarded as a sign of health and luck by the Saracatzani.

At a level basic to myth and survivals, the Greek villager is his own St. George, an embodiment of good and a force for evil. Like the saint, he partakes of the energy of life and may apply it to growth and order, nurturing the vines in the rocky soil; or he may, a moment later, direct it to destruction, plucking out, as one of the villagers did, the eye of his own brother. As Fontenrose says, "The fantasies of the myth disguise the fundamental truths of the human spirit."

The Greek peasant or shepherd has dramatic potentials for action. What response will be evoked in him depends on the uniformities demanded by the institutions and codes of his society, and on the opportunities for individual variation that arise in the interstices among sanctions, rituals, habits, and other social constants. He will also at times step outside the ordinary bounds of sanctioned acts because his ungoverned emotions compel him to and because, since he prizes his individuality so highly, he will wish to demonstrate his freedom. But in seeking to express or fulfill himself, the villager—however unexpected or idiosyncratic his acts—will usually channel his actions along routes that are at least known, if not necessarily approved, in his society.

His actions may lead him to be sporadically a saint or a devil. In either event his expressions will not be integrated within the ordinary decorum of his community. They may not be integrated within himself either. Both personality and action will appear to the observer to be inconsistent. In matters of health the outcome is usually a greater contribution to destruction than to order.

These contradictory potentials, which the Greeks recognize within themselves, are also projected upon and visualized in the external world. The physician to whom they go in the last stage of illness, expecting him to work a cure, is also considered quite capable of maliciously letting them die if they do not pay him enough. The moon, which transmits its powers to the healing herbs, must not look upon the infant lest he die. The horsebean is deadly, as the Pythagoreans implied and as its use in underworld rituals (17) shows, but it can be used as medicine when one is ill. There is the same polarity in wine: drink is good for food and mood and health, but beware the extra draughts. If a man himself does not sicken from it, it may yet lead him to murder another.

The imminent forces of the external world are perceived as having dramatic polar potentials; nevertheless, their actualization is infrequent. The facts of daily life are that people plod along quite dully for the most part. Persons or plants, doctors or drugs, the sun or the moon may have the power to do evil or good in the extreme, but one rarely sees them doing either. Drama interrupts the monotony but does not replace it.

UNCERTAINTY AND CRISIS

An outstanding consequence of both the oscillation between extremes in Greek life and the sudden eruption of extreme behavior out of the quiet daily routine is that the villager is never sure he can count on the course of future events. He has no sense of certainty, for he realizes that the drab and dreary day-to-day endeavors will inevitably be enlivened, explode, and challenge or confront him with Achilles' choice of cowardice or heroism. At one point or another he will inevitably be plunged into his own variant of tragic drama. Each life will reenact the combat myth in which the order that one has striven so hard to achieve will be threatened and perhaps torn asunder by some disruptive force, some intervening chaos. How it will happen no one can be sure. The agent may be one's neighbor, one's wife, the distrusted government official, the north wind, the raging torrent, the spirits of spring or woodland, an angry saint, a deceiving god, a sorcerer priest, or—quite possibly—one's own self.

The moment of crisis will inevitably come. Greek history is replete with crises, and so is the life of each individual Greek. The villager knows this. Foresight, as the tragedians realize, is a painful gift. The consequence of it is that the peasant who knows the world and admits its uncertainty cannot help but be anxious.* Polarity, inconsistency, and uncertainty contribute to that individual anxiety so characteristic of Greek peasants and shepherds. If there is any motif in their lives prominent enough to be singled out, it is anxiety.

People respond to anxiety in several ways. They may defend† against awareness of anxiety with measures that, while vigorous enough, are somewhat incapacitating, since they restrict the individual's freedom to function but do not reduce his actual vulnerability to threat. These are essentially neurotic responses that occur in most humans, but which will not concern us here. We are interested in a second response, which is more constructive because it attempts to eliminate the sources of anxiety, either those in the external world, or those tender spots of unreasoning vulnerability inside oneself. This response leads men to attempt to increase their actual control over future events, not by withdrawing or hiding but by extending their power so as to reduce future danger. It leads men to acquire money or full granaries, to build families where love may be found and from which security will come in old age, or to come to agreements with their neighbors so that no further boundary disputes will arise. These positive actions have become part of the fabric of village life, and although they are not always effective, many values and institutions in the community do serve, among their several functions, to build a more secure world.

There is no novelty in saying that the growth of civilization, insofar as man consciously strives for goals in building a way of life, is directed toward increasing man's power over both human and natural events in order to bring about conditions that facilitate the expression and satisfaction of human needs. The shepherd encampment or the peasant village may be conceived of as a miniature civilization—an ordered accumulation of objects and patterns of living, which, by tradition and continuing readjustment, continue over time and provide for satisfaction

* We define anxiety here in the psychiatric sense as an emotionally charged apprehension or foreboding of the future, an unpleasant mixture of hope and fear, an expectant affective response based partly on what has gone before, on what is likely to happen again, and on the vulnerability of the individual who is unable to protect himself against painful repetitions.

† Defend is used here in the psychiatric sense, implying an attempt to reduce distress by shutting out from consciousness the sources of threat while at the same time accommodating to the needs or wishes that produce threat situations.

of need and for a structured world that reduces uncertainty and ambiguity. Insofar as villagers use their powers or introduce changes that make the future more benevolent (whether by learning to use fertilizer, cooperatives, vitamins, or doctors), we may see their actions not only as foresighted, but as serving to defend against anxiety by reducing the unpredictability of life.

POWER AND MAGIC

Some of the activities in which the villagers engage emphasize order and regularity without providing them with any demonstrable increase in power over events. We are speaking now of the rituals: carefully prescribed ceremonies that are usually both sacred and interpersonal, and which involve the meticulous repetition of words and actions on each occasion of their use. Although one may infer that rituals serve to reinforce community solidarity and reaffirm values and sanctions, their intent for the villagers is magico-religious. The participants desire to secure the approval and protection of the gods. They do this by rendering the service the gods are deemed to require: the homage, sacrifices, and statements of obedient subordination that manipulate the gods by flattering them, by giving them a small gift in return for a larger one, and by vowing to give them even greater gifts for future divine performances, gifts that the supplicant secretly may intend never to give.

These rituals are necessary because the gods are conceived of as having considerably more power than humans, and because it is presumed that one can manipulate that power. But power is not limited to the gods or to men. One finds power in the sun, which not only gives warmth and life, but also causes headaches that need to be treated by magical manipulations—the bewitching of the sun. The moon also has power. One plants as it swells so that the seeds will grow in the fields or in the belly of one's wife. One does not plant in the moon's waning, for the seeds in the field will fail and one's wife will be barren. Foods and drink have power: a little is for the good and too much is for the bad. Serpents, springs, trees, and rocks also have power. They are not indifferent to man, so man cannot be indifferent to them.

In each instance, whether god, spirit, nature, plant, animal, or heavenly body, there is a power abiding that exists in relationship to man. It may be described, as it often is, in terms of intimacy, which has led some observers to speak of the harmony of the peasant with the natural world.

We think "intimacy" is overstated, and that "harmony" is much too optimistic and systematic. The attitudes of the villager are neither unified nor consistent. At best they reveal a struggle and a blending of powers among men, spirits, nature, and objects, a struggle and blending that are ordinarily preserved by accident or consent in an unsteady and potentially dangerous balance. It is a balance that can be influenced or exploited by one energy or another: a man's strong arm, a good tractor, a gift to the local doctor, a promise of a lamb to the saint, or incantations that bewitch the moon. To the extent that harmony among the forces may be inferred, that harmony is a matter of shared characteristics, of fluid conceptions that presume identities among diverse events or substances. Seeds, women, and the moon wax and wane together; the word "cancer" is the cancer itself, and the disease so summoned will, the peasant thinks, appear. As a consequence of conceived similarities in nature and in the belief that nature is responsive to manipulation, the rituals have a general function. One may try to manipulate St. George and the doctor in the same way, or seek to exorcise a headache, sterility, or a rival in love with common magical techniques. The repetitiveness and relative meticulousness of rites is apparently anxiety-reducing and satisfying in that an implicit psychological analogy is drawn from the imposition of order on small segments of the present to the imposition of order on large segments of the future.

Another aspect of these magical rituals is one that we infer and describe separately only because its elements fit our preconceptions, although no division is implicit in the act or thinking of the participants. The intent that is expressed in a ritual presumes the sympathy of word, deed, and concept: the peasants believe that by naming their wish, what they wish shall be, with the proviso that the energies of the supernaturals will be enlisted toward this end. The gods can be enlisted because the villagers know how to act to earn their goodwill—through the forms of ritual. The result is that the performance and intention of sacred rituals do reduce anxiety, whether the ritual is the elaborate Easter service, which promises rebirth; the reading of the exorcism in church to combat the frightening sorcery; or the making of the sign of the cross by a wise woman over a man with a headache. The rituals alleviate the anxiety of crisis by giving one something to do that everyone agrees is the proper thing to do, and by reducing the immediate ambiguities and dissipating nervous energy. Moreover, they reduce the chronic anxiety over future uncertainties by giving the individual a chance to manipulate the powers through the expression of magical intent and by consecrating his wish with performances that reaffirm that he, at least, is doing his assigned part in the community of men, gods, and nature.

The rituals of ancient times have been categorized by Jane Harrison (30) into two types. One type consisted of rites of "tendance" (therapea), in which men served the Olympians, using the rule of "give and be given" in an interchange where worship was mostly a matter of gods and men doing business with one another. The other type consisted of rites of riddance and aversion (apotropaic), which were conducted with fear under the rule "do and avoid" and implied a threat of dread, evil, or pollution that had to be removed, warded off, or purified. These two systems imply two kinds of powers: one group, the Olympians, was relatively well defined, with an agreed hierarchy of power and skills, benevolently disposed to man, and capable of being engaged in human-like commerce. The second group of gods was less well defined, by either characteristics or residence. Their godly hierarchy was less clear, and, at least by classical times, they were considered in terms of their threat rather than of their specialized functions.

The simple division into Olympian deities—Zeus and his retinue, and the chthonic powers of the earth, death, and the underworld—provides a rather neat parallel to the polarities in the combat myth: the heroes who stand for life and order, and the enemies who visit death and disorder. The Olympian system would stand against the chthonic system, as gods in the form of men stood against powers in the forms of snakes or spirits. As some historians would have it, the patriarchal sky gods of the Aryan invaders superseded the earth gods of the matriarchal Pelasgians, the latter becoming more and more associated with evil, just as the Olympians (as Jane Harrison suggests) became, later on, the demons of Christianity.

We have an interest in these historical practices because they extend to present times and reflect contemporary themes. There are rites of service directed toward the saints and God, and there are rites of riddance and purification directed toward the dead and sources of pollution. The service to the saints and God reflects the notion of noblesse oblige; in return for loyalty, deference, and gifts, the more powerful lords—whether saints or landowners—are expected to provide protection and support. The rituals for the dead are less patterned after human society, and imply more mystery, fear, and undefined power; one must tend the dead so they will *not* return; one must avoid the menstrual woman for fear of the damage her own power can do the god, the crops, the first bread, or the fighting man.

As in the case of the heroes and villains of combat myths, the dichotomy of good and evil powers becomes less clear upon examination. Olympians did kill, they did guide and rule the dead, and they did prove to be uncertain friends and dangerous enemies. Apollo, the god of healing, was also Apollo, the destroyer. Both Asclepius and the plague were under his direction. Consequently, one behaved in a way that would placate them, as one does today with their successors. The villager wishes to be quite sure to keep on their good side and to avoid their bad. The old chthonic powers in their manifold forms and places—snakes, grottoes, and springs—were also simultaneously benevolent and malignant; the snake was good luck as well as bad, springs and grottoes could renew the health and fortunes, and ghosts might lend aid as well as inspire terror. It is the same today, although the beliefs seem to be slowly disappearing (39).

Although we distinguish between primarily benevolent supernaturals (the saints and Panaghia of today) and the malevolent supernaturals (the more primitively cast powers of life and death), it is well to see in both systems a projection of raw human powers and emotions and more refined social concepts. The peasant interacts with the benevolent supernaturals, but his reaction to the malevolent forces is a more primitive and less socially oriented awe and fear. Yet both systems simultaneously reflect primal, individual, instincts and reactions and cultural elaborations. Both systems are complex, allowing a variety of beliefs, practices, and feelings; neither system is consistent. The peasant feels no obligation to show the kind of religious, social, or emotional consistency that fits into the logic or preconceptions of the observer who espouses neat rationality. Oscillating, coexisting, or contradicting within him, and sensitive to social demands or individual mood, are a variety of emotions and responses that may allow simultaneous or alternating expression. Rituals may be interspersed with innovation or spontaneity; saints may be approached with skepticism, humor, pragmatism, cunning, or piety; the big landowner, the lord, may be responded to with rebellion, wheedling, deference, exploitation or placating gifts; the paralysis of an arm may be simultaneously treated by home remedies, prayer, holy water, a trip to the holy shrine at Tenos, and the employment of both a magician and a medical doctor.

Contradictory attitudes of reverence toward and exploitation of the gods, nature, and humankind are a persistent feature in peasant life. As one of the peasants said, "We are of the earth; it gives us life; we are born of it and shall return. It holds us. One cannot succeed as a farmer unless he loves the soil. One *must* be religious about the earth." Yet, at the end of the day, the same man might exclaim, "We are enslaved to the earth!" and he might kick it, or brutalize it, as he kills a tree in his greed to tap all the resin. To the saint he will do homage, vowing a lamb in sacrifice—or even a bull if he is rich—protesting his devotion as he prays for the health of his son; the following month, when the son is well again, the peasant will feel slighted because the saint denies him the gift of life for a sickly newborn goat. He will light the saint no candle and will not offer the promised lamb for the healing of his son; he will say smugly in reference to the dead goat, "The saint who performs no miracles will get no service."

What if the saint had permitted the kid to live? The peasant would have bowed his head in reverence to the power and then demanded a bumper crop of beans! If a gift is given, ask for more. The clever man knows how to exploit the powers, but the wise one will do it with piety. One might say that the rural folk expect the greater powers around them to be unpredictable, and that in the face of uncertainty, they must be able to apply any technique within the range of the culturally acceptable and individually appealing that suggests itself as a means of getting what they want.

The peasants rarely reflect upon their actions or make rational evaluations of them in order to judge their results so that they might discard an unsatisfactory practice from their cultural or individual repertory. Except for a rare few, reflection costs the peasant too much effort, as Thomas and Znaniecki observed (66), and while a successful demonstration can lead to the adoption of new methods (22), the failure of a method does not lead to its abandonment. Thus one sees a complex system of ideas and techniques that becomes increasingly diverse because it accumulates more than it discards. Since it accumulates from various sources—innovations copied from strangers, ideas learned from city people, and inventions by local people—and since it retains the old elements from the bronze age through Byzantine and Turkish influences, the aggregation is hardly integrated. The villager does not mind this, for he is troubled by neither these contradictions nor his own subsequent inconsistency in action and beliefs.

3

Themes in Community Life: The Contemporary Scene

The technological developments of the twentieth century are exposing the Greek villager to a new way of life. He is learning new, more efficient methods of farming, and he has increasing opportunities to market his crops for cash returns. He has the chance to obtain an education that will provide him with an integrated method for evaluating his heritage and his thinking. He is being taught how to maintain and improve his health, and he has the opportunity for more mobility in and exchange with the world outside his immediate community. Such changes in his daily life are altering his beliefs about how to live and to preserve life, his conception of the nature of his community and its institutions, and his approach to religion and healing.

These changes in rural Greece are revolutionary, but the peasants are adapting to them with surprising ease and even eagerness. Perhaps this ease is due to their awareness of their historic role as innovators or as people in tune with urban centers, or perhaps it is due to their pride in the past and their desire to excel intellectually and artistically as their ancestors did. Most certainly their easy acceptance of change is related to the high value they place on change itself, and on education, city ways, and the pragmatic approach to living. Their accommodation to the present change seems to involve no wholesale repudiation of what has gone before or even any serious strain between those who stand for the new and those who value the old. Much of the old remains and the new is merely added to it. This pattern is most striking in healing practices, where magicians and physicians are often used simultaneously, and penicillin, holy oil, and the relics of a hero, probably a Mycenaean, may all be used in treating the same disease.

Parents who have never gone to school have children going to high school. Among the refugees of Dhadhi, four girls and two boys are enrolled in high school. In Panorio no girls go to high school, but one has had higher education, and, after graduating, served the village as its teacher. Historically, the Greeks have had deep respect for learning; consequently, educated children are becoming accepted innovators in the community, teaching their parents new skills and new ideas. Moreover, as the girls become educated, acquiring the prestige of formal learning, changes in the status of women are occurring. Men are beginning to find themselves dealing with social superiors who are women—doctors, professors, and so on. Although the new eminence of women does not lead men to regard them with any uniform respect or appreciation, it does add a dimension of admiration and respect for their new status to the love, regard, and fear accorded to women in their role as mothers.

The family is also changing. The children want to move to the city, the girls among the shepherds want to live in "real houses" instead of their beehive huts, and the sons who marry may live far away, disappointing their elderly parents and making them feel deserted. Worse, the parents may actually suffer from hunger if they are old and cannot till the soil. Although the money the son earns in the city is more cash than he would ever have in the village, it is not sufficient for a generous allotment for his parents' food and clothing. In the past, children, who provided for their parents when they became too old to provide for themselves, were the old people's insurance against hunger and want; but nowadays, because the children tend to move away when they are grown up, the parents may have to provide for themselves. They have not earned enough in their lifetime to have a reserve of cash, especially when all of the money they have earned has been spent on dowries for the daughter, city clothes for the family, or fine furnishings for their homes or the homes of their children. They complain that they are hungry, and they blame their children, especially the sons and their wives, for betraying them.

Children, for the most part, still recognize and honor their obligations to care for their parents or to send them money. But in some instances, family solidarity does break down, and some old people do not get aid from their children. Usually the village itself takes on the obligation of their support, although not always without some grumbling on the part of the givers or the necessity of begging on the part of the elderly. A more extreme solution to the parent-support problem—since there are no sufficient programs of insurance or institutions for the elderly—is to kill them. We have no evidence of this practice, but its prevalence in folklore suggests both its past occurrence and a continuing concern with the possibility. For the most part, the elderly are cared for within the family; while the old man is alive, he rules the generations that are assembled in the household and the old woman supervises the work of the women. One activity of the old woman is healing, which allots to her the role of family healer or wise woman.

Returning now to spending habits, one observes the tendency to earmark any extra income to purchase dowry items, radios, propane stoves, showy city clothes, or fancy dining room sets, a tendency which means that there is no money put aside for emergencies. Even if there is cash on hand, to be required to spend it for medicines or doctors' fees is considered a calamity. The absence of resource management in the city sense may lead to some absurd situations. For example, one woman in Panorio complained that she did not have enough money to buy food for her family, that her children were malnourished (this was later substantiated by medical examination), and that the food supplies from their land had run out. Her family had larger holdings than many, but (according to other peasants) they had not been prudent in their farming practices. While she was lamenting her family's plight, a traveling photographer came by offering gilt-framed photographs taken of her nephew's wedding for the price of $10. The woman was delighted and paid $2 down for the pictures. Then she returned to bemoaning her family's hunger and poverty.

Her complaints, which indicated her distress, were intended also as an appeal for help. We had encountered this approach before; the peasants expect persons of power or authority to give help. Government authorities, strangers, and others who have skills or powers that the villager wants available to him are given special treatment. The peasant, for example, protects himself from the possible ill will of the stranger by observing certain rites of hospitality that also serve to obligate the stranger, who is then asked for help or berated for not giving enough. This manipulation of strangers is similar to the mother's manipulation of her child; by "sacrificing herself" for her child, the mother builds up obligations in him that he can never fully discharge. If he fails to comply with a parental wish, he is reminded of his debt by reproaches like "After all I have done for you," or "My hair has turned gray for you," or traditional enjoinders like, "Treat your parents as you would have your children treat you." Parents and children, hosts and strangers, villagers and authorities, are trapped by one another in a tight net of obligations, which is constantly reinforced by emotional appeals, begging, gift giving, feeding, promises, lies, and outright demands.

Because resources are limited and life is hard, the peasants devote considerable energy to thinking about how the resources of others might be redistributed to their own advantage. Within the family certain emotional obligations exist that limit the possible methods for acquiring the wealth of a relative, just as within the village certain moral obligations limit such schemes. Both sets of sanctions depend largely on the opinion of the important people in the family or village, and on the operation of a balance of power that is not to be upset lightly. A man's opinion of himself also restricts the techniques he can use; if, for example, a man has a reputation for nobility or honesty in his dealings, his love of honor (philotimo) will make it hard for him to be anything but noble and honest in his actions, if only to maintain that shining image.

Relations with strangers, on the other hand, are governed less by sensibilities or communal feelings than by fear of the harm a stranger might cause. The distrustfulness of the villager reaches its fullest expression in his encounters with strangers; if the outsider's potential for power is ambiguous, he will be met with both "tendance" and "avoidance," an equivocalness that can be seen in the rites of hospitality. In addition, hospitality is very much a matter of pride and philotimo, and Greeks are justly known for their generosity.

Sociability is a most important feature of Greek life. Nearly all activities are group activities, and there is no Greek word for "privacy." Other feelings being equal, the Greek prefers family over friends, and friends over strangers; his contacts and social intercourse tend to be with people he already knows, and insofar as possible he deals with merchants, doctors, and other outsiders whom he knows or, preferably, to whom he is related—in order of descending preference—by blood, marriage, adoption, clan, village, or region.

The peasant's distrust of strangers, his tendency to have social contacts only with people he knows, and his awareness of who in his village has power lead him to maneuver those who have influence or "messa." When he wants a bargain, he asks a friend who is a friend of the supplier to make the purchase; when he goes to a hospital, he asks his doctor to intervene with the doctor's doctor friend in Athens. Influence channels can be brought into play by calling in, so to speak, the outstanding obligatory debts of friends or family or implicit promissory notes. A third device is to restore the tilting balance of obligation by giving gifts, a practice that, while reaffirming bonds among friends and establishing ties with strangers, may be looked upon by an observer as bribery or corruption. "Beware of Greeks bearing gifts" is as appropriate for the Greek peasant today as it was for the Greek warrior in the days of the Trojan horse.

But no matter how resourcefully the peasant may make use of the influence of others, unequal distribution of wealth and power remains; some will compare their fortunes with others and find themselves the losers. Hesiod said that this kind of strife leads to envy and competition, which sometimes makes one man work very hard to compete with another. But this competition also typically leads to a devious denial of success or fortune. The man who is successful denies it in order to forestall other people's envy, the demands of relatives, neighbors, and tax collectors, and the wrath of the gods, who also envy worldly success and strike down men who are overweeningly proud. The one who envies, on the other hand, is known to be dangerous. His admiration is a weapon in his hands, for it becomes the evil eye, a witchcraft force that visits illness or disaster upon the one admired. If the admirer himself is conscious of his power and wants to forestall it in some situation where he genuinely wishes no harm, he will perform a ritual of riddance upon himself, spitting and publicly imputing the worthlessness of that which he admires in order to protect it from the spell he might otherwise cast upon it.

The owner of an admired object is quick to recognize that to admire is to want. In order to forestall the damage of envy, and to placate the admirer and establish a more favorable balance of obligation, the owner of the admired object gives something small to avoid having to give something larger; he may offer some portion of the admired object to the person who admires it. Thus, the gypsy who admires a dress is given a coin in order to forestall her potent curses, and the government is offered acreage if the landowner fears his larger properties might be sequestered.

Whereas to admire is to want, to steal is "to like it for oneself" (67). To acquire objects that one has not worked for oneself is proper under certain circumstances; it is another means of righting the balance. If those who have do not recognize their obligation to give, an obligation that begins with parents and extends to landowners and government agencies, then the villager may simply take what he thinks he deserves. In rural areas one may take the fruits of the earth, for the belief is that nature's bounty is no man's by private right; neither landowner nor peasant is shocked to learn that tomatoes have been taken from the field, that sand has been pirated from the government beach, that wood has been cut from the regional forests, or that resin has been tapped from the trees of the landowner. Such "liking it for oneself" is not the same as stealing money, which is a shameful theft, as is stealing goods in order to sell them; the Greek rural folk, who rightfully pride themselves on their honesty, would never condone commercial theft.

"Divide it like a man and not like God," they say, asking for a fair distribution of property. "I've done many good things in life, all of them wasted. No, one cannot rely on God," said an old, almost blind shepherd woman sadly, revealing the bitter uncertainty, poverty, illness, and cruelly hard work that had been her lot in life. Her life had been no Arcadian idyll, as romantics think of the pastoral life; with all its richness it had been and was now a life of pain. While speaking to us she forgot all the pleasures of her swift-footed youth and became immersed in her despair. Her pessimism and despair were the opposite of the vaunted optimism and high hopes that occasionally excite the Greek and inevitably stimulate those who are with him.

But despair, which in its milder forms takes the form of resignation, apathy, and fatalism, and which contributes to the appearance but not to the essence of peasant moderation, is an extreme mood; it is not permanent, even though it is dominant when it is upon one and colors much of what is thought and done. It contributes to inactivity, letting things slide, and self-destructive neglect; it is epitomized in the one phrase we heard the peasants use more often than any other: "It doesn't matter" ("*then pirazi*"), which implies the English expression "Who cares?" or "It's hopeless." Any depression is a state of hopelessness that seems timeless, and the despair of the Greek is no different. It occurs whenever situations are perceived to be beyond control, when a person is overwhelmed with events beyond his changing, events that jeopardize or destroy his achievements, hopes, or ideals.

We believe this Greek sense of the tragic arises in part from their capacity for awareness, which under certain conditions allows them to experience life fully and sensitively. It is an intensity of feeling uninsulated by the reserve, distance, or modulation that protects city people who live more comfortable lives. In addition, we believe that the pessimism of the peasants is the result of defeated hopes—hopes that were generated during emotionally indulged childhoods but which are brought up short in experience in later life, hopes born also of the proud knowledge that theirs was once a great civilization that now would rise again, although it is painfully slow in its stirring.

HOPE AND DISAPPOINTMENT

The history of Greece is a history of the birth of hope, the promise or the achievement of greatness, and the fall of that which is cherished. To some extent each person's life repeats this tragic theme. Infants, males especially, are reared with tender indulgence and taught to expect parents and family always to indulge them. The child comes to feel that he is omnipotent, and on this assumption he builds his fantastic hopes. Then in early childhood, between four and six, his disillusion begins. He finds himself entrapped in the network of obligations and duties that are all the more inescapable because he has learned to strongly identify with the lives of others in his family. Reared as he has been and will be, in such a manner that being with and part of others is his only experience, he has little "private self" that allows him to objectify what is physically and emotionally his versus what is another's. The differentiation between mother and child, which Americans prize as a developmental goal, is not a Greek concern. The personality, the self of the child, is formed not just by associating with others but by merging with them. Consequently, a person is produced who not only is family oriented, highly sociable, and internally supported in crises by indivisible affiliations with others, but also is highly vulnerable to the ills, moods, and evils of others. For him being alone is loneliness, and loneliness is terror, for without the company of others, he feels that part of himself is gone.

Because of this family diffusion of personal identity, the child's strengths paradoxically are his weaknesses; he gains power and support from others, but he is also threatened by their failings; since the moods and actions of others in the family are beyond his control, he experiences anxiety. In the family situation the child learns early the techniques of exploitation, manipulation, and ritual control, which are later used in the uncertain world outside the family to deal with fellow villagers, strangers, nature, and the supernaturals. Insofar as these techniques fail to work—and they must fail in a volatile, uncertain, and changing society—anxiety continues but shades off, after repeated disillusionment and defeat, into despair.

It is difficult to gain perspective when so much of one's view of oneself is in terms of others, and when one's moods and hopes may have originated with another. It is the more difficult as one oscillates between poles of optimism and depression, and as one has within oneself not an integrated set of values and viewpoints, but an inherited and eclectic collection of contradictory notions and ways for judging the world.

The world of the villager is changing rapidly: both the old and the young villagers desire change, seeing in it the way to prestige, easier living, and the gratification of hopes and needs. But social alterations do not contribute stability to perspectives. Because change occurs—schools are built, electricity comes to town, hospitals are opened, money jingles in the American-made blue jeans—the hopes of villagers are renewed, but without providing them any standard that limits their expectations. Because the changes are so slow—"The other village has a road but we do not," or "Athenians are rich and we are poor"—the more fantastic hopes are repeatedly dashed, and pessimism sweeps into their place.

The omnipresence of despair helps account for otherwise inexplicable behavior: surrendering in battles that are nearly won just because the rumor of defeat is heard; failing to complain against an outrage because one knew of another who, previously outraged, complained to no avail; or giving up on a prescribed medication for tuberculosis after the tenth day of treatment because the sickness is still present. Many are the surrenders, the voiceless protests, the presumptions of failure without a fair test, and the lack of follow-through because depression has sapped a person's energy and hope.

LEISURE AND INERTIA

Inactivity should not be confused with depression. Being idle is an accepted way to spend one's time. In each village there are times of idleness, and in the nearby towns men who have dedicated their lives to inertness sit in the coffee house, letting their wives till the soil, or beg for food from the rich landowner, to support their wine, coffee, and talking habits. Looking at these townsmen, or watching the village children sit quietly staring over the fields for hours at a time, one suspects that rural life produces an infinite capacity for boredom. The origins of inertia may be fatigue, malnutrition, and an outlook more Eastern than Western; activity is not a goal in itself. The villager may say of the Americans or Athenians, who rush about, that their activity is a form of madness.

The pleasures of leisure may also account for some of the reluctance to embark on new ventures, especially when the activities are not linked to immediate satisfactions of pride or family prestige, achievement of status, or acquisition of admired city goods. A large rock rests in the middle of a road. One man could displace it with a moment's effort, making the path of animals and vehicles smoother and safer. But no one will move that rock, and until a heavy truck slams it off the road, it will remain there. In this case "Why bother?" means that effort which does not gain oneself or one's family either pride or goods is of no consequence. "Let us sit and watch the rock; perhaps someone else will move it." The same inertia can be observed in the person with a chronic cough who refuses to take the trouble of going to town to see the doctor. "It doesn't hurt; why spend the money?"

Inertia and boredom have as their corollary passivity: "I will wait; someone else will do it for me," or "Let that fool build us a schoolhouse; when it is done is time enough for us to become interested," or "Here, you are rich; you can do this for us." Such words are not uncommon. They imply the view that the man who gets others to do things for him is clever, and that if a man can arrange to be given things without incurring reciprocal obligations, he has achieved the pinnacle of cunning. Thus there is in the passivity of the peasant an element that is used to exploit the strong.

THE RESPONSE TO POWER

For their part, the ones who are powerful are not above treating the weak with brutality. One sees this in the relations between father and child, husband and wife, employer and worker, and public authority and lesser citizens. A tavern owner, kind enough in his relations with equals, will relish beating the town drunk; a little boy will be chased down and tortured with wrestling holds by a smiling older lad. In neither case will any bystander interfere. The only restrictions on the exercise of power are counterforces or anticipated bad consequences. The abuse of power, when counterforces are not present, may be held in check by a sense of noblesse oblige among the sophisticated or mitigated by elaborate networks of obligation and relationships that bind the interests of the strong to those of the weak. For the weak themselves, who lack powerful friends or the reasonable hope of retaliation, the best defense against abuse by the powerful appears to be in vigilant distrust, deference, and clever interpersonal manipulation.

Any new contact between the weak and the strong is fraught with danger for the weak. Exploitation and betrayal are the feared consequences. Knowing that the weak expect the strong to treat them badly, one can better understand the ritual hospitality with which the stranger of unknown power is received. One can also understand the covert rebelliousness or hostility of son toward father, of citizen toward police, or, occasionally, of nonpaying patients toward government doctors. In the consequences of the abuse of power one can also see additional reasons for the depression, pessimism, and fatalism occurring among village folk.

The freedom of the strong to act as they see fit, however, does lead to constructive acts. The wealthy landowner may embark on projects of vision and munificence that aid the people of his district. The benevolent despot may work with unfettered imagination and generosity to bring about radical improvements in social conditions, public health, or agricultural methods. This freedom to use power (because of the absence of complex encumbering institutions of the sort one finds in the bureaucratic committees, compromising legislatures, or opinion-sensitive officials in the United States) does result in dramatic reforms and experiments at village, regional, and national levels in Greece. The drawback is that the accomplishments do not alter the passivity of the peasant who accepts largesse without any feeling of involvement or any motivation to support projects that are under way. Without local involvement and responsibility, the best-intentioned social experiments—the gift of a tractor, a village well, or a medical center—may fail to be effective because those who use it may fail to work for its maintenance or efficiency.

Now a social revolution is at hand that will result in more equitable distribution of powers in which the rural folk will share. No longer will they have to remain subjects, for they will come to participate in decisions affecting themselves. To achieve the benefits of distributed political power they will have to unlearn the current pattern of passivity. Their new learning must extend to matters of nutrition, sanitation, and health.

FUNCTIONAL INDIFFERENCE

Indifference is another term that can be applied to villager reactions, especially in situations where the observer might expect more dramatic responses. We distinguish between indifference as a purposeful or functional response and the boredom or inertia previously discussed. Indifference is a way of denying the importance of what has been said or done, and it serves to release the individual from an implicit demand that he take some action. In effect, he looks the other way when something happens that he knows ought to move him but when he prefers to avoid involvement. He knows the cost of involvement only too well, having had a lifetime of experience with aroused passions, betrayal, anxiety, and disaster. Indifference allows the villager to indulge his cowardice without admitting it; he can watch a thief steal without sounding the alarm and risking his own safety. Indifference can also allow him to let his brother's illness go unattended, or to deny his fears about his own chronic stomachache.

One denies the facts or withdraws from them when the consequences of admission are too costly, either in anxiety, in danger, or in futility. These are the conditions under which the villager's capacity for sensitivity and awareness is numbed or excluded. There are several features of the villager's life that account for the prevalence of the indifference. One is self-interest, which dictates that a situation that does not affect one's personal or family values or welfare does not require commitment. In a social world oriented to the family rather than to the larger commonweal, disavowal through indifference is an easy resolution of conflicts between one's morality and the events one observes.

A second feature that predisposes the villager to indifference is the nature of philotimo. Philotimo is the Greek male's self-image and pride. It is a sense of honor and worth. Cultural dictates determine its concerns. One's philotimo demands revenge for any dishonor or insult. It is enhanced by one's hospitality, physical vigor, and ready response to family obligations. The Homeric Greek had philotimo—Achilles' conduct was shaped by it: he was blameless and unsurpassed in strength, vain-glorious boasting, and revengeful sulking when his honor was at stake. It was not for the sake of a mere woman, Briseis, that he nearly killed his own war chief, Agamemnon, who demanded that Briseis be turned over to him. Nor did Agamemnon want Briseis as a bed-companion—he did not touch her—and it was unlikely that Achilles had done so before him, since a man going into battle would avoid the pollution emanating from woman, which weakens manly valor. Honor was the issue that set Achilles against his companion in arms. He was to be robbed of his prize of honor (geras); that the prize was a woman was secondary. The measure of Homer's heroes and of Greek men today is taken at each encounter. Each instant may bring social or physical injury; each moment may bestow the passing glory of having bested the other and asserted one's manly honor. So serious is the matter of philotimo that a man—villager or sophisticated Athenian—may fall physically ill when he loses face because of shortcomings or failures that can be neither denied, rationalized, nor avenged.

A man must love honor to count among his fellows. To defend his honor, to let no insult go unavenged, entails anxiety, for each day he may be required to prove himself a brave man. The tests are not infrequent, for ridicule is common and is delivered with gusto by groups of men who, in the relative safety of their numbers, thoroughly enjoy the challenge to the solitary butt.

Because of this frequent need to be philotimous, and the strain and anxiety such a challenge causes, indifference becomes an alternative to the all-or-none response dilemma. One does not heed nor even hear the insult. One ignores the challenge. This indifference seems to be a necessary alternative because it is difficult for men to sustain themselves at peak intensity for long. By being indifferent, a man can sustain the philotimous myth of himself without being put to the test. "Every man thinks he is a king," said one employer, whereas the reality was that in his organization they were abused as pawns. By maintaining defensive perceptions that shut out the challenges, conflict with others and threats to the philotimo are eliminated.

Along with their defensive indifference, the villagers seem to be inattentive to time and to certain kinds of space and activity. Perhaps events outside a person's own affairs are not considered relevant; at any rate, much that goes on escapes their notice. A Red Cross health-education team can spend a whole day working with some villagers and the others will never note the team's presence. Inattentiveness may extend to failure to see increasing physical disability in a relative or even in one's own bodily functions, especially those that do not directly interfere with role activity (being a mother, a farmer, etc.) or valued aspects of the self (complexion, vigor, potency or fertility, humor, appetite, and sociability).

MEN AND WOMEN

Within the family poor communication contributes to ignorance of what goes on. We were struck by how little husband and wife tell each other. We found it unwise, for example, to arrange a medical appointment for a husband through his wife, for as often as not she would not tell him about the arrangements and sometimes she would not even mention the possibility of a medical examination to him. The phenomenon appears to be part of a more general social distance between the sexes in matters of sharing, sympathy, and life orientation.

Marriages are arranged by the parents. The choice of spouse is designed to provide for the future of the bride through selection of a husband whose occupation promises economic security and, ideally, an upward movement in status. Educated or town men are prized. The choice of a bride is likewise made with an eye to improving the status of the husband and providing, through the dowry, for the couple's initial welfare and competitive position. Naturally, richly dowered brides command the husbands with the greatest status potential.

The dowry is a great financial burden for most village families. Its accumulation takes place over many years. The entire family takes pride in giving a good dowry, but that satisfaction is offset by the hardship that accompanies its provision. The anticipation of the dowry problem is one reason why baby girls often are not welcomed into the family.

Neither the arranged marriages nor the attitudes of men and women toward one another facilitate affectionate ties or easy communication between the sexes. Exceptions occur. It sometimes happens that a boy and girl marry for love, and there are a few marriages in which love and respect have developed over the years. But even in these cases women are subordinate to men.

In the villages beatings and abuse of women are not uncommon. The women usually work harder than the men and are required to be forbearing in the face of harsh treatment. The woman's training in showing deference to the more kingly male begins at the age of two or three, and one sees the seriousness, reliability, tolerance, and sometimes the depression of very young girls. The role that gives the village woman most satisfaction is mothering. She nurtures both children and husband, and performs both household and farming-shepherding tasks. The very extensiveness of women's work means that men neither deign, nor are able, to fend for themselves in any domestic matter; not even when a man is only slightly sick will he think of looking after himself. The woman finds satisfaction in the man's dependence on her; she may console herself for her hard work with the thought that "All men are children," and that "Women can be alone but men cannot." The women's belief that the male is the psychologically weaker sex may or may not be correct, but in any case, the belief heightens their own self-esteem and provides rationalizations for times when men make them suffer. They feel a sense of confidence as they expand their power through maternal nurturing.

Men enjoy their dominant role. There seems to be no question that a man's leisure evening hours in the taverna are easier and more pleasurable than the woman's evening work of cleaning up, weaving, processing food, spinning, and so forth. The philotimo of the male waxes glossy in the glory of his household domain, and he rears his girl children to be subordinate. This rearing process, which is usually accomplished with a mixture of indulgence, demands, and not a little physical punishment, may also lead to particular strains between man and wife as the daughters reach puberty. For then, as in rural areas elsewhere (72), incest may sometimes be practiced, with the father, or both the father and the brothers, exploiting the growing girl. Neither the daughter nor the wife is allowed a protest in this male empire, but one infers that there is some increased strain on husband-wife relations in those few families where incest is actually practiced. While one need not presume any psychological ill-effects on either party, the social consequences may require drastic action. If the girl should become pregnant, there is the likelihood that either her brothers or her father will kill her. This kind of killing—an "honor" killing in which the honor of the family is protected by the murder of the girl who has soiled it—is usually explained by claiming that she had become pregnant by an outsider with whom a marriage could not be arranged. In point of fact, there is evidence that honor killings—none had occurred for several years in any of the villages we studied—may reflect the father's or brother's jealousy of an outsider who has usurped family domain, or the blame-throwing reaction in its extreme, as the biological father kills the badge of his own shame.

Such a custom leads us to consider, as one aspect of male-female relations, the more fearful characteristics of women. Women are considered dangerous. They have powers over men and nature that can lure, weaken, and destroy. Part of this belief is a convenience; if a man rapes a woman, it is her fault for having tempted him; thus the village women take care to dress so as to conceal their bodies. A girl child is a constant danger; she is a prime source of potential family dishonor, inherent in the sexual temptation that she carries and the possibility of her public shame. By tempting successfully she weakens and pollutes man, and the tabernacle as well, should she enter it.

The male concern with woman's powers, her fertility, maternal control, devious evil, temptress role, power-annihilating sex, menstrual or postpartum blood—which weakens even God, so that she may not appear in church during these times—is reminiscent of the attitudes held in ancient times toward the chthonic deities. Today these ideas are rarely expressed publicly, and for many villagers the ritual practices could not be related to the superstructure of beliefs about potency and fear. Nevertheless, ordinary behavior and comment reflect tiny segments of what the observer interprets as a larger theme: menstrual women are feared and isolated; they do not visit the sick; one does not have intercourse before taking holy communion; women are "devils," and their "shame"—specifically, their genitals—must be concealed and periodically purified, as in ancient times, by ritual bathing in the sea. As can be imagined, such attitudes toward female sexuality pose special problems for women patients who are seeing male physicians.

4

Illness

In this chapter we turn to the specific matters of reported illness and health behavior. We shall present statistics obtained from the morbidity survey and medical examinations conducted in each community and information gathered in the formal interviews and spontaneous discussions that accompanied the survey. The methodological cautions raised in the Introduction apply especially to this chapter, and it may be well to repeat that what villagers say and what they do are not necessarily the same, and that their reporting of illness is likely to underestimate the extent of the community's actual morbidity experience.

THE EXTENT OF ILLNESS

All but two of the 50 families in Dhadhi, or 96 per cent, reported that someone in the family had been sick during the preceding 12 months; 63 per cent of all persons in the village were said to have been ill at some time during this period. In Panorio all but four of the 29 families, or 86 per cent, said someone had been sick during the past year; 43 per cent of the community's inhabitants were said to have been ill. In the Saracatzani encampment all eight families reported that someone had been ill in each household; 55 per cent of the persons in the encampment were said to have been sick at one time or another during the preceding year. Table 1 presents the most frequently reported illnesses in the three communities.

Upper respiratory infections (for the most part, colds and flu), which had the highest general incidence rate, were probably underreported by about 600 per cent (see Appendix I). The considerably different inci-

TABLE 1

Most Frequently Reported Illnesses Occurring in a 12-Month Period in All Three Communities

(Ranked by incidence per 1,000)

	Dhadhi		Panorio		Saracatzani	
Illness Group	Incidence	Rank	Incidence	Rank	Incidence	Rank
Upper respiratory (acute)	275	1	302	1	141	2
Stomach and intestinal disorders	140	2	42	4	190	1
Joint ailments	110	3	67	2	119	3
Cardiovascular	104	4	18	5	24	4
Allergies	60	5	0	0	0	0
Genito-urinary	55	6	50	3	24	4
Smallpox vaccination reaction	5	7	50	3	0	0

dence rates for stomach and intestinal disorders (including acute disturbances, ulcers, and amoebic dysentery), which ranked second overall, perhaps reflect the differences of awareness, recollection, and definition of illness,* as well as the more marked differences of environment, nutrition, resistance, and risk, which contribute to differential morbidity rates.

Notice should be taken of the high incidence of actual illness reactions to smallpox vaccinations, which were given at the time of a threatened epidemic in Greece. One must presume that a larger proportion of Panorio villagers had never been vaccinated in childhood; or that they were more anxious and therefore experienced an anticipatory bad side effect; or that there was some defect in the vaccine or in the method of administering it. Inquiry revealed that more villagers were vaccinated in Panorio than in the other two communities during the nationwide campaign; this may account, at least in part, for the high rate of illness reactions. Sixty-nine per cent of the Panorio families reported members who had been vaccinated recently, whereas the figure was only 25 per cent in Dhadhi and 20 per cent among the Saracatzani. An estimated corrected incidence rate for illness reactions to the smallpox vaccine in Panorio is 94 per 1,000. (This is a rate based on only those vaccinated and reporting reactions, and is different from the rate of reported

* For example, some villagers said they would not classify a head cold, a headache, or a low fever as an illness because these were an ordinary part of living and did not deserve special identification.

reactions in Table 2, which includes unvaccinated as well as vaccinated persons.)

Although the illnesses listed in Table 1 are the ones most commonly reported, there are others, which, though infrequently named, play an important role in health beliefs and actual village health. Among the serious infectious diseases reported as having occurred within the past year are meningitis, scrofula, trachoma, malaria, salmonella, and amoebiasis. Interesting for the very infrequency with which they were reported are the usual childhood diseases. In the three communities, with a total of 36 children ten years of age or under, only three children were said to have had measles, mumps, chicken pox, whooping cough, or similar diseases. Our observations suggest that the villagers take no note of these diseases and make no effort to take temperatures, put the child in bed, or keep him home from school. These illnesses are considered "compulsory diseases"; children are purposely exposed to them. Since they are considered necessary, are hardly noticed, and do not arouse any concern, we assume that marked underreporting accounts for the low incidence quoted. Other illnesses that are infrequently reported are the folk diseases, about which, the villagers say, doctors have no knowledge. The morbidity survey elicited reports of "the wandering navel," "the waist out-of-place," "the bubble," "heaviness," and "the white or

TABLE 2

Persons Reported Ill by Age and Sex Compared with Total Population in Each Group

Age and Sex	Dhadhi Ill/Population	Dhadhi Per Cent	Panorio Ill/Population	Panorio Per Cent	Saracatzani Ill/Population	Saracatzani Per Cent
Children						
5 and under	11/24	(46)	4/9	(45)	0/1	(0)
6 through 9	14/22	(64)	6/10	(60)	1/6	(17)
10 to 18	10/18	(56)	5/14	(36)	2/9	(22)
Adults, 19–59						
Males	31/47	(66)	14/43	(33)	6/11	(55)
Females	30/51	(59)	19/31	(62)	12/13	(93)
Adults, 60 and over						
Males	13/18	(73)	2/7	(29)	0/0	(0)
Females	16/19	(85)	3/9	(33)	2/2	(100)
Total	125/199	(63)	53/123	(43)	23/42	(55)

the shadow on the eye."* These and others will be discussed in later chapters.

Table 2 shows the number of persons reported ill by age and sex during the 12 months preceding the interview. These data show that men and women, adults, and children are all reported as having suffered illness. The reader is reminded that we do not interpret these reports as reflecting actual illness prevalence; instead, we assume that the illnesses recalled are those that were significant for the family (with the exception of concealed diseases and those conditions which, although illnesses by medical standards, are not defined as such by the villagers).

In Dhadhi 45 per cent of all reported illnesses were chronic ones having lasted more than one year. Another 14 per cent had lasted from two to 11 months. In Panorio 36 per cent of the illnesses reported were chronic, having lasted one year or more, and 17 per cent had lasted from two to 11 months. Among the Saracatzani 87 per cent of reported illnesses were chronic, having lasted over one year. Reports of short-term illnesses were approximately equal in number among the communities; these illnesses were categorized according to duration: one to three days, four to seven days, one to two weeks, and three weeks to one month.

The greater number of chronic illnesses reported by the Saracatzani may be interpreted as reflecting more accurate recall, and probably also the greater social significance of long-lasting illnesses. Table 1 would not support the inference that the actual incidence of chronic disease was any greater among the Saracatzani. It can be seen that they reported about half as many acute respiratory illnesses as did the other two communities but about the same number of chronic diseases. The medical-examination findings reported later in this chapter do not reveal any remarkable reduction in acute morbidity among the Saracatzani, nor do they suggest a chronic-disease level dramatically higher than that found in the other two villages. We conclude that the Saracatzani dramatically underreported short-term illnesses and the disabilities attendant upon them.

Disability defined by bed rest was found to accompany 54 per cent of all reported illnesses in Dhadhi. The median time spent in bed for all those ailments that required bed rest was from four to seven days.

* It is difficult to define these terms because their use in folk diagnosis is by no means precise. We have attempted to do this, however, in the Glossary, pp. 251–52, which lists Greek words and specialized terms used in the text.

Nearly ten per cent of illnesses requiring bed rest caused the patient to spend from one to 11 months in bed. The 92 bed-rest illnesses reported in the community for the preceding year amounted to 3,534 days in bed, or an average of 38 days in bed for each illness requiring bed rest. For the village as a whole, this means an average of 18 days of in-bed disability for each inhabitant of Dhadhi. In Panorio, 68 per cent of the reported illnesses required bed rest. Among these 50 cases, the median disability period was from one to two weeks. The average time in bed per disabling illness was 31 days, or 1,550 days in bed. For the village as a whole, this means an average of more than 12 days in bed during the preceding 12 months for each person in Panorio. Among the Saracatzani 35 per cent of the illnesses required bed rest. For these 13 cases the median period in bed was four to seven days, an average of 731 days in bed or 56 days in bed per disabling bed-rest illness. Extended to the entire shepherd community, this means an average of 17 days in bed per person for the preceding 12 months.

When those families who had reported some illness during the previous year were askcd if anyone was still sick, 72 per cent of the families in Dhadhi said someone was ill at the moment, as did 59 per cent of the Panorio families and 100 per cent of the Saracatzani. For each family reporting current illness the average was about two members ill.

RESULTS OF PHYSICAL EXAMINATIONS

Under the preceding head we have provided a brief summary of the reports of illness that we secured in home interviews with families. The question is: How accurate were these self-assessments of disease? And to what extent did a family's report—which sometimes consisted of a recital of a previous doctor's diagnosis—correspond to the opinion of an examining public-health doctor?

Out of three villages with a total population of 371 persons, 261 persons, or 70 per cent, presented themselves for examination by the team of public-health doctors, nurses, and laboratory technicians. The examinations, which were held in the public school of each village, began with recording the medical history of each patient. About one-third of the villagers indicated to the public-health physicians that they had had one of the usual childhood diseases, and one-fourth said they had had upper respiratory infections. About one-quarter of the people also said they had had malaria. Other serious illnesses that villagers recalled included typhoid fever, amoebic dysentery, tuberculosis and scrofula, diabetes, jaundice and hepatitis, pneumonia, meningitis, diphtheria, undulant fever, scarlet fever, and cataracts and eye diseases.

A comparison of these data with the data elicited during our household interviews reveals that the villagers underreported in their answers to the physician's questions. For example, six times more joint disease was reported in the household survey as having occurred during the preceding year than was reported in the medical histories as having occurred during the entire lifetimes of the patients. The household sample shows as much stomach and intestinal disorder during the preceding year as the medical history shows for lifetimes. More cardiovascular disorders, including hypertension, were also reported in household interviews than in the medical histories, and only a few more acute respiratory illnesses were reported over a lifetime in the medical histories than in the household interviews for the preceding 12 months.

The reasons for such underreporting are no doubt the same as for other interview results. Greek doctors often complain that patients conceal information, especially about diseases considered shameful: tuberculosis, epilepsy, venereal disease, and mental disorders. In addition, villagers are afraid even to mention these dreadful ailments, thinking that the name will invoke the illness. If they have had tuberculosis they might call it "a kind of a cold," and if there has been cancer in the family they might refer to it as "the exorcised," adding, "Let the empty hour hear it." When they speak of death, they first say, "Let only the empty hour listen," thereby expressing the magical intent that death himself, old Charon, will not hear them and be summoned.

Another reason for the inaccuracy of the medical histories is that the physician often tended to reduce the flow of information. Sometimes he used loaded questions, such as, "You don't have any difficulties, do you?," expecting, and usually getting, the negative reply. Or he hurried on to the next question before the patient had time to answer. The short time available for compiling the medical history of each patient also limited the amount of probing the physician did; signs of hesitancy or worry in the patient were not pursued. One patient complained after his examination, saying, "When the doctor asked me, I told him something was wrong with my stomach, so he wrote down 'stomach trouble' on the paper. But why did he believe me? Why didn't he examine me to find out?" The circumstances in the field, the interview habits of the physicians, and the psychological barriers to full expression by the patients all contributed to the inadequacies in the medical histories.

The symptoms or difficulties that the patient reported having at the time he saw the doctor are called the "present complaint." Table 3 shows the complaints most frequently reported to the examining physician.

The villagers were asked by the physicians what medications prescribed by a doctor they were taking. Derived rates for prevalence or current use of medications were 139 per 1,000 for Dhadhi, 99 per 1,000 for Panorio, and 135 per 1,000 for the Saracatzani. "Doctor's medicines" were most commonly used for the treatment of stomach and intestinal disorders (rate of use, 22 per 1,000) and for high blood pressure (22 per 1,000). We found that a few patients were taking medicine for bronchitis, diabetes, prostate troubles, malaria, flu, and liver trouble. It is interesting that a few of these ailments were not mentioned to the doctor when he was taking the histories. It may be that this phenomenon is related to Cartwright's finding that once an illness has been diagnosed or treated, its symptoms cease to be mentioned in response to an interviewer's questions (14).

The first set of findings from the clinical examinations was based on height and weight measurements and the general physical examinations. In the total population of children under 17 who were examined, ten children out of 63 from Dhadhi were diagnosed as severely undernourished, or dystrophic. Seven children out of 27 from Panorio were dystrophic, and three were underweight. One child of the Saracatzani was diagnosed as dystrophic and four as underweight. Combining dystrophy and underweight into the category of malnutrition, prevalence rates for children under 17 were 159 per 1,000 in Dhadhi, 370 per 1,000 in Panorio, and 333 per 1,000 among the Saracatzani.

TABLE 3

Complaints Most Frequently Reported to Examining Physician

Complaint	Cases (N = 272)	Incidence per 1,000
Joint pains	41	150
Headache and/or dizziness	33	126
Gastrointestinal and stomach disorders, including abdominal pain	32	117
Loss of appetite	13	48
Chest pain, difficulty in breathing, expectorating pus	12	44

The most common disorders revealed by the physical examinations were hearing loss or defect (58 cases, or a prevalence rate of 215 per 1,000); loss of visual acuity, as measured by the Snellen chart (59 cases, or a prevalence rate of 217 per 1,000; tonsillitis, quinsy, overgrown tonsils, or acute sore throat (41 cases, or a prevalence rate of 150 per 1,000); indigestion or gastritis (23 cases, or a prevalence rate of 85 per 1,000); hernias and ruptures (15 cases, or a prevalence rate of 55 per 1,000); and hypertension (13 cases, or a prevalence rate of 48 per 1,000). Table 4 presents the final diagnoses of the physician when he had completed the histories and clinical examinations.

Certain diagnoses were made by hospital pathologists on the basis of routine blood and urine examinations, which were conducted as part of the physical examinations. Among 124 Dhadhi residents cooperating in the blood study, 14 case of anemia and one case of borderline anemia were found; this is a prevalence rate for nonspecific anemia of 113 per 1,000. Among the 86 Panorio residents who gave blood, six were found to be anemic, four of these borderline cases; this is an anemia prevalence rate of 23 per 1,000 (excluding borderline cases). Among the 32 Saracatzani giving blood, none was diagnosed as anemic. Urine examination of 104 Dhadhi residents indicated that none was diabetic. Out of 85 urine specimens from Panorio, one indicated diabetes, another contained blood, and a third contained pus; for the latter two cases no diagnoses were established. None of the 27 Saracatzani giving urine was diagnosed as diabetic, although one shepherd was taking medication for diabetes. In Dhadhi two persons were found to be suffering from hereditary enzyme lack (G6PD).* These cases were investigated in a special study conducted in association with our investigation.

Thirty-three Dhadhi patients were advised to see a specialist or have further tests and laboratory studies in Athens. Thirty-one Panorio residents and seven Saracatzani were similarly advised. In sum, one out of every four patients needed further specialized examination or treatment recommendations. For many of these patients no final diagnosis was entered by the public-health physicians because further study was necessary to establish a final diagnosis. Chest X ray for suspected tuberculosis was the most frequent procedure recommended (16 cases). Ophthalmologists, ear-nose-and-throat specialists, and surgeons were the most frequently recommended specialists.

* Glucose-6 phosphate dehydrogenase.

TABLE 4

Final Diagnoses in the Three Communities

Diagnoses	Dhadhi (N = 144)	Panorio (N = 91)	Saracatzani (N = 37)
No disease present	41	35	11
Dystrophy, malnutrition	16	10	3
Rheumatopathology, arthritis	16	3	4
High blood pressure	12	3	2
Tonsilitis, overgrown tonsils	10	1	6
Bronchitis, asthmatic bronchitis	8	3	0
Hernia, rupture	6	4	0
Ulcers	6	0	0
Gastritis	5	4	2
Gout	3	5	0
Low blood pressure	3	2	1
Scrofula	3	2	0
Enteritis	3	0	0
Sciatica	2	0	1
Chronic appendictis	1	2	0
Rhinopharyngitis	1	0	2
Neuropathology, autonomic disorder	1	0	1
Tachycardia	0	0	1
Headache	1	1	0
Ptosis of stomach, intestines	2	0	1
Neonatal jaundice	1	0	0
Abrasions from accident	1	0	0
Eczema	1	0	0
Allergic response to wasp sting	1	0	0
Varicose veins	1	0	0
Liver sensitivity	1	0	0
Hemiplegia	1	0	0
Amoebic dysentery	0	1	0
Skin disease, not specified	0	1	0
Constipation from prostate operation	0	1	0
Kidney disease	0	1	0
Neurosis	1	0	0
Otitis-rhinitis	2	0	0
Tumor, skin	1	0	0
Gallstones	1	0	0
Pleurisy	1	0	0
Adenoiditis	0	2	0
Kidney stones	0	2	0
Allergy	1	1	0
Malaria	0	1	0
Muscle pain	0	1	0
Diabetes	1	0	0
Cataract	1	0	0
Cardiac failure	1	1	0

We had hoped to be able to include diagnoses made by these further studies in this report, but, as Chapter 7 reveals, most of the patients failed to go to Athens for the recommended work. For those few who did go, we were able to learn the results in only five cases: three blood studies were positive for thalassemia, and two chest X rays were negative for pathology.

FAMILY REPORTS COMPARED WITH PHYSICIANS' FINDINGS

In comparing the family reports of illness with the doctors' clinical findings, one must remember that the family interviews and the medical examinations did not take place at the same time. The family interviews had extended over a period beginning three to four months before the medical examinations were made. The last family morbidity interviews took place the day before the medical examination. To an unknown extent, therefore, the differences between reported morbidity and medically observed morbidity may be due to changes in actual health status from springtime, when family interviews began, to late summer, when medical examinations were held.

Table 5 presents both the number of persons in each village reported by families as "currently ill" and the number of persons found by the physical examination (excluding laboratory findings that were not included in findings by the examining doctor) to have diagnosable diseases or disorders. In view of the inconsistencies between Tables 4 and 5, we have counted as "ill" all those who had some finding of pathology and who did not receive from the physician a final diagnosis of "no disease present" or "healthy."* The reader is asked to keep in mind that the difference in population size between the interview group and the examination group represents the loss of persons who were unwilling to participate in examinations. In Table 5 the proportion of persons reporting disorders in each community is expressed as a percentage of that community's total population.

Before discussing these singular and consistent differences between the family reports and the medical examinations, it should be noted that those persons appearing for examination were self-selected and may differ in their morbidity from those who did not appear. It is possible, therefore, that the sicker persons did go to the physicians; on the other

* Villagers with loss of visual or auditory acuity but with no other illness are, by this method, included in the healthy sector.

TABLE 5

Persons Currently Ill: Family Vs. Medical Reports

Community	Family Report		Medical Report	
	Ill/Population	Per Cent	Ill/Population	Per Cent
Dhadhi	73/201	(36)	103/144	(72)
Panorio	27/125	(22)	56/91	(62)
Saracatzani	15/42	(36)	25/37	(68)

hand, it is just as possible that those who had less illness came for examinations. Morris has shown that in an English population he studied, those who were most willing to cooperate in medical examinations had less illness than those who refused to cooperate (44). We found that working-age men made up the largest proportion of the villagers that were not examined: 54 per cent of this group did not appear. The second largest group that did not cooperate was composed of men and women 70 years of age or older. Twenty per cent of this group failed to participate. Disproportionate rates of illness among the working men is unlikely, but in the aged group this likelihood is great.

Assuming that the bias in the self-selection for medical examinations was not great enough to alter our present findings, and that seasonal variations in morbidity were not so great that illness was much more common in summer than in spring, we conclude that there is considerable underreporting of morbidity in the family interview as contrasted with clinical findings. The interviews revealed that 31 per cent of all villagers were said to be currently ill. We assume that this figure represents the villagers' own perceptions of the state of their health. But the medical examinations showed that 68 per cent of all examined villagers had a current illness or disability. The discrepancy may be even greater, because the findings of many suspected cases of illness, in which recommendations for further specialized examinations were made, were not included in the figure reached by the routine medical examinations.

These differences are consistent from one community to the next, although Panorio reported less disability in the family interview and this was corroborated by the medical field examinations.* However, if the

* At the very beginning of our study in Panorio, when we were introduced to the villagers at the community assembly in the coffee house, they speculated that the reason

Panorio people had gone to Athens for specialists' examinations as they were advised to do (in proportions higher than other villages, since 34 per cent were so advised as compared with 23 per cent among Dhadhi patients and 19 per cent among the Saracatzani), it is likely that their final diagnoses based on laboratory studies would have shown a proportionately greater increase in morbidity prevalence rate.

It is useful to compare the reports of chronic illness in the family interviews with the clinical findings of physicians. Table 6 presents this information, and also gives the occurrence of chronic illness as reported by the patients in the medical histories. The data are combined for all three communities.

Table 6 shows that nearly half of the family reports of joint diseases were not sustained by medical examination. One physician's comment on this was that "they think they have rheumatism, but it is only muscle pain from fatigue and overwork." It is difficult, on the other hand, to account for the discrepancy between the reports of cardiac disease and the medical findings. Cardiac complaints may reflect the misinterpretation of chest pain as heart trouble (whereas it is more likely to be tuberculosis, if the number of suspected cases recommended for X ray is indicative of TB prevalence). But there is also the possibility of medical underreporting of cardiac disease. Table 6 also shows possible overreporting by families of allergies and kidney and bladder ailments; family underreporting is most dramatic in the case of tuberculosis, especially if any of the 16 suspected cases did have the disease. Underreporting of tuberculosis is, however, understandable, because of its gradual onset, the difficulty of distinguishing it from fatigue or from cough due to benign respiratory ailments, and the belief of the villagers that it is a shameful disease.

The widespread idea that some diseases are shameful is one cause of the underreporting. Tuberculosis, venereal disease, epilepsy, cancer, skin blemishes, pocked skin, and mental disorders are all considered shameful. Occasionally hernia, crippling ailments, visible goiter in a girl, the presence of lice, or the absence of menstruation are also considered

---

we had chosen Panorio was that they were all so healthy. "Ninety per cent of us are healthy," they said, adding they had had only one death in the last three years. At that time they said with pride that their good health was due to hard work, clean air, good water, and (with ironic humor about their hunger) "not eating too much." Family reports seem to reflect this proud village ideal of being in better health than other villages.

TABLE 6

Comparison of Household Interview and Clinical Findings on Chronic Illness

Chronic Diseases and Disabilities (Partial List)	Household Interview (N = 362)	Medical History Interview (N = 272)	Diagnosed by Physician (N = 272)
Rheumatism, arthritis, gout	35	41	21
High blood pressure	13	0	17
Cardiovascular other than hypertension	10	5	3
Hernia, rupture	4	1	10[a]
Diabetes	2	0	1
Allergies	11	1	2
Cataract	3	0	1[b]
Kidney and bladder ailments (presumed chronic)	18	1	1[c]
Gallbladder disorder	2	0	2
Tuberculosis, scrofula	2	0	5[d]
Venereal disease	0	1	0
Epilepsy	1	0	0

[a] 15 reported in systems review not mentioned in final diagnosis.
[b] 2 reported in systems review.
[c] 3 reported in systems review.
[d] Plus 16 recommendations of chest X ray for suspected tuberculosis.

threats to honor or reputation. A few villagers contend that any disease is shameful, for regardless of what it is, they say, people will ridicule the sick person, and may even refuse to enter into a marriage with someone whose family has a history of any kind of serious disease.

We discovered several explanations for the villagers' belief that illness is a threat to esteem. One is that strength and beauty have esthetic value; families are proud of their members who are strong or beautiful. Weakness, physical disability, or ugliness, as in a crippled or scarred person or a hairless man, are thought of as shortcomings—just as they were in classical times. Since the affliction of one member reflects on the entire family (shame, as well as honor, is possessed by the family as a group and depends on the conduct and qualities of each of its individual members), any illness that produces ugliness or disability is a threat to family esteem.

Unfavorable public opinion is a menace not only to the family honor, and to the chances for marriage of its sons and daughters, but also to the individual philotimo. The presence of a shameful illness in the family threatens especially the father's philotimo because he has the power and responsibility in the family. Therefore, especially in more traditional families, definite efforts are made to conceal the blemish of reprehensible ills.

In addition to the threat to family esteem that illness causes, the shame of social rejection or isolation occurs when other members of the community fear the illness. For example, in one village a man had epilepsy. His family concealed this fact as best they could, but, of course, his seizures became known among the villagers. Nevertheless, the family would not inform the investigators of his illness. Only after another villager told us about it, in the greatest confidence, were we able to learn about it from the family. When we indicated our awareness in casual conversation, the family was no longer defensive and they expressed their despair. They believed the disease was hereditary, and the mother of the family feared that it was causing sickness in the other children. Examination proved their sicknesses to be no more than minor disorders; nevertheless, these diseases, believed to be the consequences of the epilepsy, were viewed as signs of weakness, vulnerability, and hereditary taint.

One villager was psychotic and had been hospitalized. Only after the interviewers had known the family for a long time was the fact revealed by the family's request for help in getting further diagnostic work. They indicated that they had concealed the schizophrenia from others; it was very important to them that no one else know. Nevertheless, others in the village did know, and although they were kind enough to the sick person, it is not certain that they did not devalue his family.

Another villager had been orphaned as a child. His father had died of the "bad pimple" (anthrax), and his mother of an unknown ailment. Two older brothers had died from tuberculosis when they were young. As an orphan lad, this villager had been well treated by others, but he was nevertheless considered to be a deviant bearing a hereditary taint. His family's history of disease was considered proof of a blood impurity; as a consequence, no family in the village or region would even consider allowing their daughter to marry the last remaining male of a vulnerable and "polluted" line.

A disease or disability that prevents an individual from fulfilling a fundamental biological or social role is also shameful. For example, a person who is crippled and dependent is called "only half a man."

Since the bearing and rearing of children is a primary source of satisfaction and a social duty for parents, any condition that interferes with childbearing is a terrible blow. Women who do not menstruate or who are barren or men who are impotent or sterile suffer greatly. Their conditions are shameful; their philotimo can be tragically challenged, and they may be disdained as deviants in the community.

It is an easy step from feeling deeply depreciated to searching for an ego-salving excuse or a way to shift the blame. Sexual or reproductive failures are often attributed to the malice of others, specifically to the binding curse of sorcerers; epilepsy and madness are usually attributed to the devil or demons. These various explanations will be discussed in Chapter 9.

The impact of the community's rejection is not so great that a person cannot adjust to his predicament if he has strong character and can develop a somewhat independent philosophy. For example, in one village there was an older man who said he had learned when he was 20 that he was sterile and that he had taken two steps to solve his problem. First, he had a Voronoff operation, in which monkey testicles were implanted; and second, he slept with his reputedly barren fiancée for some time to make sure that she would be satisfied with him as a husband even if the surgery failed. It did fail, but both husband and wife report a satisfactory marriage. Obviously, harmonious matches of biological deviants can be made if each of the partners has sufficient psychological strength. These two described their way of life as one where they minded their own business and disregarded community attitudes that labeled the husband as ugly because he was hairless.

Certain illnesses imply a hereditary pollution and threaten the philotimo, but there are more aspects than shame to the attitude toward these ailments. Sicknesses that are considered contagious are feared, and villagers may avoid or isolate the sick person. In a society where social contact and sociability are highly valued and being alone is fearful and unnatural, the sick person is especially sensitive to this rejection. The following account, given by an old wise woman, illustrates the handling of one case:

"Tuberculosis used to be one of the shameful illnesses. People don't want to talk about it if they have it because they are afraid that they can't get married. My playmate when I was a girl in Asia Minor had an illness like that, although it wasn't tuberculosis. One day she developed a boil or something like that on her behind, near the base of her spine. From that day on she disappeared. We used to go to her house to call her to play, and her mother told us she had gone to another town. Of course, this was a lie; the mother was simply afraid that someone would hear about the disease and the girl wouldn't marry. She used to bring the doctor in at night, secretly. Maybe it was cancer.

"Years passed, and it was only when the disaster at Smyrna happened [the Turks burned the town] that we saw her again. Her waist was almost broken in two, and they carried her out on a stretcher. When they took her to the island where her brothers were, they couldn't accept her into the house or the town, so they made a hut for her outside the village. Her mother used to visit her there and care for her until she was all rotten and they let her die. And the flies ate her.

"So you should know that if there is an illness you should always go to a doctor. If you have such a worry, find someone you trust, a member of your family like your mother or sister, and discuss it. Never keep it a secret to yourself, for, as people say, 'A hidden pain is never cured.' That girl could have been saved if she had been taken to a doctor and hospital in time."

The possibility of cure is now altering the concept of shameful ills. An illness that can be cured is demonstrably not hereditary, at least as the villagers see it; and the sufferer is not condemned to lifelong disability and isolation. Demonstrations of cure are also slowly changing folk ideas of the causes of illness. One of the most important ways for public-health authorities to overcome the many damaging social customs and dangerous health practices associated with the "shameful" ills of tuberculosis, cancer, epilepsy, skin disease, and psychosis is to educate villagers to the nonhereditary nature of most of these disorders and their treatability.

Venereal disease is shameful because it is evidence of bad conduct. Sexual intercourse is not considered "immoral" in the Protestant sense, but for single women it represents the squandering of virginal assets that should be saved for use in the marriage negotiations. For both men and women venereal disease demonstrates pollution, a ritually and spiritually impure state that may prohibit the sufferer from successfully seeking the help or blessing of the saints or God. Venereal disease is also shameful because it is contagious and therefore feared. The sick person is a carrier of physical danger as well as pollution. Venereal disease may also be considered evidence of abnormality. The sufferer is said to have proved that he has abnormal sexual desires; that is, he desires sexual congress with strangers—city prostitutes and the like—from whom he can expect only ill fortune.

One consequence of all shameful ills is the expectation that one's enemies will rejoice in them: one's misfortune is the pleasure of his foes. Thus the shameful ill is a public weakness that gives a social victory to one's enemies. Since community life, outwardly stable and harmonious, seethes with envy and distrust, there is no shortage of enemies who will triumph over the family burdened with such diseases.

OTHER REASONS FOR UNDERREPORTING

The underreporting of illness reflects not only faulty recollection and, perhaps, the unwillingness to recognize and admit shameful ills, but also the definitions people have of illness: the bodily conditions that occur before the decision is made that one is, in fact, "sick." Definition of illness is an extremely important consideration, not only for morbidity reporting, but for the recognition of illness that is necessarily the first step toward treatment (10). The signs and symptoms the villagers use in deciding whether they are sick, in the order of frequency of their mention to us, are the following: (1) pains, aches, fever, chills, and malaise; (2) dizziness, weakness, heaviness, and tiredness; (3) inability to work, to eat, to walk, or to talk; (4) irritability and bad humor; (5) depression, loss of "courage," and worry; (6) paleness and yellow skin color; (7) vomiting, cough, dry lips, sleepy eyes, muscle stiffness, injuries and wounds.

This is in many ways a comprehensive list. It includes the common symptoms of acute illness, the mental states associated with the progress of disease, and the definition of illness in terms of loss of function in the significant areas of working, eating, and socializing. The list does not omit any common symptoms except, perhaps, those associated with lesser alterations in body processes or appearance, such as diarrhea, constipation, or visible tissue changes (skin eruptions, boils, changes in warts or moles). Coughing was mentioned by only one family as a sign of illness. While other respondents may have overlooked this sign, it is likely that some do not consider it as evidence of an illness. Indeed, it is possible that tuberculosis, lung cancer, and bronchial disturbances are not suspected early simply because coughing does not often lead to the definition of illness.

In the course of the medical examinations, one woman was observed by the public-health nurse, during chest measurements, to have an inverted nipple and a painful hard mass in the breast. Asked about this, the woman said she had noticed it, but had not regarded it as dangerous. She added that she had no intention of bringing it to the attention of the doctor, not only because she did not connect it with illness but also because it would be immodest to allow the physician to examine her breasts. Several women, during the course of the study, referred to their edematous ankles and spoke of having shortness of breath. They were aware of these conditions, but in the absence of pain or other conceptual criteria for defining illness, they did not see any need for medical attention.

These are a few illustrations of how criteria for illness definition shaped self-diagnosis. These concepts, of course, depend on health information. The peasants' limited hygienic and medical knowledge in recognizing illness is by no means peculiar to Greece. Studies in England and the United States demonstrate that large numbers of these populations are not aware of bodily conditions that indicate illness (38, 44, 69).

When the signs of illness are vague or when several diagnostic interpretations are possible, the villagers may decide on the name of the illness by seeing how it responds to various kinds of treatment. When a person complains of dizziness, high blood pressure is suspected, and he is given several different herbs. The type of high blood pressure diagnosed will depend on the particular herb that eliminates the symptom. One wise woman said she knows 14 different kinds of high blood pressure, each of which is diagnosed by its specific treatability with herbs.

A method for simultaneous diagnosis and treatment is used for complaints suspected of having been caused by the evil eye: Olive oil is dropped in a cup of water. If it separates or "disappears," the illness is due to the evil eye and the symptoms will simultaneously "disappear." If the oil does not separate, then the illness has not been caused by the evil eye, and the patient and the wise woman must consider other healing steps.

Among peasants and shepherds there are several conditions that are believed to be beneficial even though the local people realize that physicians disagree with them. Most of the families interviewed said that they considered boils and pimples to be beneficial. (There is a common belief that each boil saves one from another disease.) There are various explanations; some say that boils "clear the blood"; others believe they are necessary for maturation in adolescence; and still others say that boils in a young man are simply a sign that he should marry. Implicit in the latter notion is the belief that semen is constantly manufactured and that unless the male has intercourse he will suffer physically, that "the semen will build up in his bone marrow." The boil is a sign of this, and we may speculate that semen once might have been thought to escape through the boil. The following account shows how, according to the peasants' belief, any undesirable element (an emotion, a spirit, etc.) that enters the body can be ejected through pimples: "I have very 'hard' blood and it won't keep anything bad in my body. When I worry I get all kinds of pimples and things and then the 'bad' leaves me. When I was young I had pimples as big as the head of a pin. . . . They were signs of worry that my blood wouldn't keep and so it threw them right out of my body."

One family in Panorio and one Saracatzani family spoke of intestinal worms as beneficial, saying that they were necessary for digestion. It is interesting to note that this concept occurs elsewhere; for example, Ackerknecht (1) found it among the Thonga of Africa. A few villagers speak of discharges from the ears or swellings as being beneficial. One family noted that induced burns are good for infections. Using what is apparently a form of cauterization, they have adopted (whether on empirical or on magical grounds) a traumatic treatment method that can have beneficial results.

Along with the widespread belief that pimples are beneficial, however, the villagers also recognize a dangerous ailment called "the bad pimple." In Panorio especially there are stories of sudden deaths from this pimple, described as having a black core, which is undoubtedly anthrax. Nevertheless it is considered an ailment the doctors do not know and, as such, would not, in the past at least, have been brought to the attention of a physician if it had occurred; nor is it probable that it would be reported in a medical history.

There are a number of other conditions that the peasants and shepherds think physicians have not heard of, do not recognize, and do not know how to treat. Nearly all villagers agreed on the lack of awareness among physicians of the following conditions: the wandering navel, the evil eye, jaundice, the waist out-of-place, the anemopyroma (apparently facial erysipelas), the korakiasma, the "white" or "shadow of the fly" in the eye, bewitchment by the sun, the moon, or the stars, "the spleen," the "exorcised," and all those sicknesses caused by sorcery or the exotika. In addition, some of these conditions were defined in terms of their treatment. For example, the doctors are said not to recognize ailments (most of them caused by sorcery or the evil eye) that can be cured by the xemetrima (the ritual words used in healing, as in a spell or incantation), by cutting (here they refer to jaundice, which is "cured" by cutting the labial frenum, a membrane connecting the inside of the upper lip to the gum), and by cupping and rubbing. These conditions are ordinarily diagnosed and treated by folk healers, and are not spontaneously brought up in a discussion with a physician.

Any public-health program would need to educate villagers to the advantages of medical treatment for the "beneficial" illnesses, as well as for the illnesses that "doctors do not know." Such a program, of course, assumes the availability and efficacy of medical care within the region.

5

Birth, Abortion, and Death

This chapter reports data from a subsample consisting of half the families in Panorio and Dhadhi and all of the Saracatzani families on births, miscarriages and abortions, neonatal deaths (six-month term to 10 days old), and infant and child deaths (age 10 days to 21 years). It also presents relevant information about beliefs and practices in regard to birth, abortion, and death.

Table 7 presents statistical data on the reported births, abortions, and deaths. In the table women are listed in two different groups, one beyond the childbearing age, which has been arbitrarily set at age 48, and the other of childbearing age, 47 or under. For women in the older age group the rate of giving birth was nearly double the rate reported for women of childbearing age. Some of this difference is undoubtedly due to the fact that women in the younger age group have not yet had all the babies they might be expected to have. However, it may also reflect a tendency toward smaller families among the younger women, a trend reported in other areas of Greece by Friedl (26) and by Vasilios (71). Such an interpretation is supported by the relatively high average age of the women in the childbearing-age sample. In Dhadhi the average age is 33 years; in Panorio it is 34 years; and among the Saracatzani it is 42 years. It seems unlikely that the members of this group will double the present number of their children in the childbearing years still ahead of them.

Dramatic changes in public health have occurred over the last 25 years. Life expectancy in Greece has increased by about 15 years since 1930 (71). In the period from 1926 to 1930 it was 49.1 years for males

TABLE 7

Reported Miscarriages, Abortions, Births, Neonatal Deaths, and Infant-Child Deaths for Two Age-Groups of Mothers in All Three Communities

Category	Dhadhi 48 and Over (N = 13)	Dhadhi 47 and Under (N = 17)	Panorio 48 and Over (N = 9)	Panorio 47 and Under (N = 6)	Saracatzani 48 and Over (N = 3)	Saracatzani 47 and Under (N = 4)
Miscarriages, abortions	5	7	3	3	1	1
Births	43	42	54	15	18	14
Average birthrate per mother	3.3	2.5	6.0	2.5	6.0	3.5
Neonatal deaths[a]	4 (10)	2 (.5)	3 (.6)	0	2 (11)	2 (11)
Infant-child deaths[a]	13 (30)	2 (.5)	7 (13)	3 (20)	4 (22)	0

[a] Per cent of births given in parentheses.

and 50.9 years for females; from 1955 to 1959 it was 63.9 years for males and 68.9 years for females. These nationwide improvements are reflected in the village data. Reported neonatal deaths have declined in both Dhadhi and Panorio, although they have not declined among the Saracatzani.

Table 8 presents estimates of the number of children to reach age 21 for the two groups of women. The estimates are rough at best because few of the children of women in the younger group have reached age 21, and some might die before reaching maturity. Assuming, however, that none of the children now living will die before age 21, a comparison can be drawn.

If the estimates in the table are reliable, comparison of older and younger age groups in Dhadhi allows the inference that birth rates and infant-child mortality rates there are declining at about the same pace. This results in about the same number of children living to age 21 for the younger women as for the older women. In Panorio and among the Saracatzani, on the other hand, the birth rate appears to be falling more rapidly than the infant-child mortality rate. If this is correct, there is an actual decline in the number of children reaching age 21. The inference is consistent with other data on the declining family size and population in rural Greece.

The limitations that apply to statistical data reported by the villagers are particularly applicable to Table 8. Some older women said they

TABLE 8

Estimates of Number of Children Reaching Age 21 for Two Age-Groups of Mothers in All Three Communities

Averages	Dhadhi 48 and Over	Dhadhi 47 and Under	Panorio 48 and Over	Panorio 47 and Under	Saracatzani 48 and Over	Saracatzani 47 and Under
Birthrate per mother	3.3	2.5	6.0	2.5	6.0	3.5
Number of children lost by age 21	1.3	0.3	1.1	0.5	2.0	0.5
Number of children per mother reaching age 21	2.0	2.2	4.9	2.0	4.0	3.0

could not remember how many babies they had had. Some women could not remember how many of their children had died, and many appeared reluctant or unable to recall abortions or miscarriages. It was not unusual for women to speak of "abortions" at nine months term. Throughout the interviews and especially during spontaneous discussions, it was apparent that induced abortions were common, although they were not likely to be reported. For many village women the induced abortion represents the only means of birth control. Another population-control device used in the rural areas is infanticide, but we can give no accurate estimate of its incidence. At times in the discussions with the village women, it was difficult to distinguish between abortions, "nine-month abortions," still-births, and infanticide.

Another limitation of the data in Table 8 is that they include no sampling of unmarried women. Although it is not unusual for an engaged girl to have intercourse with her fiancé, pregnancy is dreaded, and when it does occur, the couple must marry. If they do not, the girl's life is in danger. Those women who spoke of their pregnancies before marriage did so freely and without embarrassment because each of them had married and legitimized her own and the infant's status. Anyone who had experienced a pregnancy that had not been legitimized would most certainly not have told us about it. Under such circumstances, the pregnant girl might seek an abortion, flee to a large city, or even kill the infant at birth.

Although infanticide is abhorrent to the villagers, they nevertheless indicate that there are circumstances in which it is justified. They usually justify it in terms of the belief that unbaptized babies are creatures somewhat apart from the human family, and that some of them may in fact be "devils" rather than human beings. There are many stories of infants' being discovered to be demons. The person who makes such a discovery, the villagers believe, has every right to protect himself from the supernatural dangers and the unnaturalness presented by the presence of a demon. One does this by invoking apotropaic spells and rituals to dispel the creature, by visiting a magician who can exorcise the demon nature of the infant and retain a human in its place, or by throwing the demon away or killing it.

There is no question that fear and awe are associated with all infants, and that in spite of their obvious helplessness, they are thought to possess supernatural power. Indeed, in some places in Greece, as in the Slavic lands, unbaptized infants are referred to as "monsters" having the power of some of the old chthonic deities. Although these fears no doubt reflect the awe of the villager for the miraculous power of reproduction, these same beliefs provide a rationalization for destroying an infant. If a parent decides that the baby is not really human but a demon, he does right by destroying it. The characteristics of demon infants, as described by the villagers, fall into two categories. One set of demon traits is obviously supernatural: the creature takes various shapes and forms, it is found at the crossroads, it turns into a bat and flies away, and when picked up it grows and grows and grows, taking the form of a snake, a monster, or a demon.* The other set of traits is quite human but unpleasant or worrisome: a demon infant is very thin (this illustrates the importance attached to feeding and fatness in this culture), cries a lot, or "is a lot of trouble." A baby that is too troublesome may cause a parent to suspect him of being literally a "monster" and to seek a solution by fair means—magical or foul. Although these beliefs are widespread, to our knowledge, no infanticide (except one in an adjacent area) occurred during our stay in the region.

ABORTION, BIRTH CONTROL, AND SOCIAL CIRCUMSTANCE

Birth control is an important concern of the village women. Most of them in the child-bearing age group expressed a desire to limit the size

* Sometimes the infant's devilish nature is directly attributed to his origins, as when a villager says that "Diavolos has the unwanted children," or that the sins of the parents will torture their children. These beliefs reveal the externalization of blame that has been so characteristic of Greek culture for thousands of years: the creation of a personified or anthropomorphized source—a Ker or Diavolos—to account for events or behavior that man does not want to acknowledge as his own or even as human.

of their families. Their reasons were closely related to their life circumstances; the poverty of most families dictates the wisdom of a planned family. Each child represents an additional financial drain on resources that are already quite limited. Pregnancy and early child-care requirements reduce the time available to women for working in the fields and thus further reduce crop and income potentials. Pregnancy is also a fatiguing experience, which either temporarily or chronically reduces the mother's physical capacity for productive work.

A female infant is a more likely candidate for a "nine-month abortion" than a male, partly because of the low position of the female within the family and the potential threat to family honor that she represents. Whereas boy children are a source of great pride and future security to both parents, girl children are welcomed more by the mother as companions and co-workers. Especially if there are other children, a new girl-child may not delight the father. He is disappointed not only because a girl can earn less money than a boy and her dowry is costly, but also because her sexual attractiveness could destroy the family honor.

Any or all of the foregoing factors—economic strain, physical difficulty, social demand, and the threat to honor—can make any pregnancy a potential disaster. But the reasons against having children are counterbalanced, of course, by strong reasons for having them. These include the natural desire for offspring; the importance of the family as the basic social unit and the desire to expand that unit; the great pride taken in having children (it is a compliment to and a demonstration of the virility of the male and the womanhood of the female); the high value on children by all members of the community; the value of children as helpers in work; the role of children in giving a purpose to their parents' life (the purpose of marriage is to have children and the reason for subsequent work and effort is to rear them); and the strong expectation that the children will support the parents when the parents reach old age, thus providing them with an essential social and economic security, without which they would face isolation and starvation. The existence of these two opposing forces suggests that each pregnancy must be weighed carefully. If there are several children already, one must assume that the likelihood of a decision against further children is increased.

Knowledge of modern contraceptives among the villagers is not evident, and it is not certain that an educational program for their use would be allowed because the Orthodox Church is officially opposed to them. Our impression is, however, that the official position on contraceptives is not a matter of great importance to the local priests. At any rate, except for lemon juice and quinine douches, no modern contraceptives are available for the women in the region. As a result, the responsibility for contraception rests mainly with the men.

When the men fail to use contraceptives and the douches fail to be effective, abortion will be considered as a birth-control device. It is difficult to say how often abortion occurs. Married women, as we said earlier, are reluctant to discuss this subject. The only other estimates available are from local midwives. The midwife is likely to have some information about abortions, both because she has long acquaintance with the women in regard to childbirth and baby care, and because in many areas she is the one who performs the abortion. Abortions are also performed by physicians, although in the communities under study it was said that local physicians did not do abortions; it was necessary to go to Athens for the operation.

One midwife informant in our area, a bright and well-trained woman, indicated that abortions were heavily relied upon as a birth-control method. To induce abortions, women sit in hot water, and use suppositories and local medicines. According to the midwife, some women have aborted by these methods as many as 30 times; she estimates that the average village woman will be aborted by midwives or doctors ten to 15 times during her child-bearing years.* Abortion by a physician is expensive, costing from $10 to $175. If it is done by a professor of obstetrics in a university, it will cost about $230. There is no set price because the operation and its arrangements are secret.

Another costly consequence of intercourse for the unmarried girl, aside from abortion or childbirth, is the necessity of re-establishing her virginity. Virginity is a requirement for marriage. An exception is made only when a marriage has been contracted and intercourse occurs with the fiancé during the engagement period. For all other girls it is crucial that previous sexual experience be concealed. One attempt at concealment is the surgical reconstruction of the hymen. Two surgical procedures are employed. One is a temporary restoration that is done imme-

* One knowledgeable priest estimated that 20 per cent of all infants conceived are intentionally aborted or killed at birth; an additional 5 per cent die or are aborted (in late term) accidentally. He doubted that child killing occurred.

diately prior to the marriage and may cost $35; the second method is a permanent restoration, which is much more expensive. The amount of money required for either operation poses a considerable financial strain on the girl.

REPORTED CAUSE OF DEATH FOR NEONATES, INFANTS, AND CHILDREN

Mothers in the three communities were asked why their babies and children had died. The causes of death in each village are given in Table 9, which reflects a number of folk beliefs about the dangers to life and health. It also implies the absence of proper medical care for the mother and baby, for where the cause is not described medically or where death is due to a presumably preventable disorder (enteritis, umbilical hemor-

TABLE 9

Reported Cause of Death for Babies and Children (Including Stillborn)

Cause of Death	Dhadhi (N = 199)	Panorio (N = 123)	Saracatzani (N = 42)
Flu, bronchitis	3	1	0
The "bad hour"	0	5	0
Evil Eye	2	1	0
Meningitis	2	0	0
Enteritis	2	1	0
Fright[a]	1	0	0
Strangled by soldiers	1	0	0
"Sickness" during pregnancy	1	2	1
Measles	0	1	0
Whooping cough	1	0	0
Paralyzed during delivery	0	1	0
Malaria	1	0	0
Lack of care during mother's absence	1	0	0
Dengue fever	0	0	1
Contaminated water (probably typhoid)	1	0	0
Mother unable to eat something she wanted	1	0	0
Milk spoiled by new pregnancy	0	1	0
Poisoned by horsebeans	0	0	1
Anemia	0	0	1
"Infection"	0	0	1
Blood coming from navel	0	0	1
Unknown	5	0	1

[a] Stillbirth blamed on mother's being frightened by guerrilla soldiers and Germans.

rhage, measles, malaria, etc.), it must be assumed that proper medical care was either not available or not used. The two deaths attributed to war reflect the tragedies that have afflicted the Greeks during this century.

The "bad hour" is the cause of death most frequently reported, but it is reported only by mothers in Panorio, where it accounted for more than a third of all infant and child deaths. The "bad hour" is a term applied to a group of spirits or demons which usually take the form and shape of an animal such as a goat or a black dog, and which are members of the larger class of supernaturals called "exotika." Although the "bad hour" may also be used figuratively to describe the occurrence of a dreadful event, its usage even then implies a specifically supernatural cause. The "bad hour" is blamed for certain illnesses that afflict adults as well as children. It will be discussed further in Chapter 9.

The evil eye is believed to be another fairly frequent cause of death for babies. This belief, which was as widespread in the Mediterranean world in antiquity as it is now (20), holds that the eyes of a person may express or direct a power that can harm people or objects. The evil eye is one of the most important belief systems used to explain illness, death, or misfortune.

Three other beliefs are reflected in Table 9. One is that horsebeans are dangerous to infants and children, causing sickness or death.* This belief is widely held, and its historical importance is attested to by the doctrine of Pythagoras that proscribed the eating of beans. Another belief is that a pregnant woman's food cravings must be satisfied or there will be dire consequences. This belief, which is not restricted to Greece, is associated in the villages with the custom allowing any pregnant woman to demand of another householder any food she smells or desires. It is the householder's obligation to appease this demand. If he does not give the pregnant woman what she asks, any miscarriage that ensues will be his responsibility as an inhospitable householder. Dorothy Lee (40) reports that when a pregnant woman passes by, the householder who sees her may rush to her to proffer whatever food may

* The disease "favism" is associated with a genetic defect in which the enzyme glucose-6 phosphate dehydrogenase is lacking. Some people from the Mediterranean basin lack this enzyme, and when they eat the favabean (horsebean) a hemolytic crisis occurs. We may presume that this reaction had been observed in ancient times and is reflected in the Pythagorean writings. The villagers' belief that horsebeans are dangerous to all people, especially to children, is erroneous.

be cooking so that he will not be responsible—or blamed—for any misfortune to the unborn child. Lee states further that the pregnant woman who is offered food cannot refuse it for fear of placing both her baby and her host in jeopardy.

This custom reflects several features of Greek culture. One of these is the great importance attached to food and the ritualization of feeding practices as a means of maintaining health and procreative power. The custom also reflects the belief in the special vulnerability of pregnant women, who must be protected from untoward events. Avoidance and tendance rituals (the latter illustrated by the food offering) provide community support for the expectant mother. Pregnancy is surrounded by a good deal of magic for warding off harm emanating from the evil eye, the exotika, the stars or the moon, and sorcery. Diet is important in these protective regimens.

A third belief, that a nursing infant dies if another pregnancy occurs, appears in Table 9. This idea may symbolize and provide arguments for the belief that frequent pregnancies are undesirable. Thus, expressed as a health rule or a "valid" reason, the desire of most of the village women to have few pregnancies has community support and sanction.

The belief that a baby was born dead because its mother was frightened by the guerrillas and Germans during World War II reflects the emphasis generally given by villagers to the emotional causes of illness. (We do not discount the possibility that war-strife induced miscarriages, or that wartime conditions in the region, especially during the period of starvation and guerrilla fighting after World War II, might actually have caused miscarriages.)

DEATH

Householders were asked if anyone in the household had died within the 12 months preceding the interview. One death was reported among the total of 362 persons in the three communities. Another person died as the study was being completed.

In discussing death, the people of Panorio and Saracatzani take special pride in saying that their people die when they are very, very old, and in remarking on how vigorous they are until the end. One man spoke of his father, who could eat four pounds of bread at one meal until the day he died; another recalled the "levendi" who at 80 could dance and drink all night. One old shepherd spoke happily of his 79 years, attributing his sprightliness and longevity to the fact that he had never seen a doctor or taken a pill, "not even aspirin." These examples illustrate the pride people take in a long, full life.

The villagers personify death as Charon; for some, death is a spirit one can talk to and bargain with about the day one must die. There are no consistent sets of beliefs about events after death; contemporary beliefs seem to be an aggregation of ideas drawn from the ancient world and, to a considerably lesser degree, from Christianity. Hades, the underworld, is rarely visualized as heaven or hell; it is more often a Homeric abode of shades. Nevertheless, a villager will practice good works for the "good of his soul," which, it is assumed, will benefit thereby in an afterlife. Some villagers tell stories suggesting reincarnation; others worry about ghosts. There are tales of the vrikolakes, or revenants, whose improperly buried bodies did not decay, so that the soul cannot rest and must wander about to annoy the living. Vestiges of the ancient cults of heroes are still apparent, and chthonic rites from cults of the dead have been incorporated into Christian practice.

6

Treatment Activities

This chapter introduces information on treatment activities. Later chapters will take up again the subject of treatment in relation to the villagers' participation and cooperation in medical care (Chapter 7), local and regional medical facilities (Chapter 14), and folk healing and folkhealers (Chapters 7 and 13). We begin with the interview data about reported actions for the treatment of the specific illnesses suffered during the previous year, as discussed by families during the morbidity survey.

We found that folk (traditional) healing activities in the form of home "practika" or home remedies were the most frequently reported kinds of treatment. These home remedies are either known to the patient and his family or have been recommended by neighbors or folkhealers (in the latter case most often one of the village wise women). Examination by a local doctor was the second most frequent step in medical care. A visit to a doctor in Athens ranked third as a treatment method and seeking admission to an Athenian hospital was ranked fourth. A "formal" visit to a folkhealer, either a skilled man or a wise woman in the village or a famous healer living elsewhere, was least often reported.* A total of 191 contacts with medical doctors was reported for the population of 362 persons in the three villages, or an average of about 0.5 visits per person per year.†

* The infrequency of reported visits to a folkhealer is, we believe, partly due to the informal nature of these contacts, so that an unknown number of "visits" is represented in the use of the recommended home manipulations and medicaments. As discussed in Chapter 9, there was underreporting in the morbidity survey of the use of traditional healing beliefs and practices.

† This figure may be contrasted with the United States average of about 4.5 visits per year.

HOSPITALIZATION

The nearest hospital for the villagers is in Athens. The trip there must be made by bus (one bus a day goes to Dhadhi, four buses pass through Panorio, the Saracatzani encampment is near the bus route), and usually takes from one and a half to two hours.* Some villagers have never been to Athens, and some have been only a few kilometers out of the village. For most of them Athens is a strange and frightening place, albeit an exciting one. For all of them the trip there is an extraordinary event.

The two local physicians who live in Doxario are general practitioners who practice in the English rather than the American fashion. That is, they do not undertake any specialty work or surgery, not even the setting of fractures. When any problem requires consultation, specialty skills, or hospitalization, they refer the patient to a hospital clinic in Athens. (Most public clinics are parts of hospitals.) The governmental midwife in Doxario, a trained public-health nurse, does not deliver babies herself, except in emergencies. She supervises prenatal and postnatal care, but she recommends that all her patients—including the women of Panorio, Dhadhi, and the Saracatzani, whom she occasionally sees—go to the hospital for their deliveries.

The villagers tend to rank physicians in terms of the urban status system. The professor at the university is considered to be a better doctor than the local practitioner. The city physician attached to a hospital is also considered a better doctor than his country counterpart. The peasant is impressed by the title and badges of status. Given free choice, he would seek medical care in the city, and even though he hates to pay a doctor's fee, he is convinced that the more the doctor charges, the greater are his fame, skill, and efficacy.

These are some of the factors that influence the peasants and shepherds in using physicians and hospitals. Local physicians, the nurse-midwife, and their own judgment influence them to go to Athens, but practical problems and psychological barriers influence them to stay home. When asked what disadvantages are associated with going to the hospital, the villagers set forth the following (in order of frequency):

(1) Expense (including hospital costs, transportation costs, and loss of income from not working).

* No villager owns an automobile, although several jointly own trucks. The cow-owners cooperative in Dhadhi also owns several trucks, but these are not supposed to be used for private purposes.

(2) Discrimination in quality of medical care; doctors and nurses are said to make poor people wait or to give them less medical attention.

(3) Lack of confidence in hospital facilities; hospitals are said to be crowded, to lack equipment, and to be dirty.

(4) Dissatisfaction with the human side of hospital care; doctors and nurses are said to be impolite and unkind; they do not tell patients enough about their illnesses and treatments.

(5) Visiting privileges are too restricted; one cannot see one's child or family except at special times that are inconvenient for villagers who must travel some distance to get to the city.

(6) Fear is aroused by the idea of hospitalization; the villagers feel insecure about leaving home and the village.

(7) Distance; transportation to the hospital presents very serious difficulties for the sick and their families.

(8) Neglect; villagers think that hospital patients are not given good care.

(9) Government hospitals are inferior; the villagers believe that doctors and nurses require gifts or tips to provide service approaching that given private patients.

(10) Influence is necessary to secure attention; without influence with officials, admission or treatment is hampered.

In answering the question about the disadvantages of hospitalization, only 7 per cent of the villagers indicated that there were no disadvantages. The majority of families expressed some concern about the expense, distance, administrative practices, or quality of care in the hospital. The barrier these worries raise was shown in Chapter 4, when we observed how many patients failed to go to Athens for the further examination and treatment recommended by the public-health team. Nevertheless, when pain and anxiety make it imperative that the person no longer deny or neglect his illness, he will go to Athens for care. Table 10 presents hospitalization data for all the village families for the 24 months preceding the interview.

Not all villagers must pay for hospitalization. Many apply for and receive indigency papers that authorize free care in the third-class facilities of public hospitals. To be classified as indigent, they must appear before a social worker and prove that their family cash income is less than $384 per year,* a ceiling that, according to our estimates,

* There are three categories of indigency. The first category includes families whose annual cash incomes are less than $4.00 per family member per month, the second those whose cash incomes are less than $6.00 per family member per month, and the third those whose incomes are less than $8.00 per family member per month. No level of indigency or eligibility for free care can be established unless the family's annual cash income is less than $384.

TABLE 10

Hospitalization Information

Hospitalization Data	Dhadhi (N = 199)	Panorio (N = 123)	Saracat-zani (N = 42)
Number of different persons hospitalized during 24 months	26	17	3
Number of admissions during 24 months	30	17	3
Duration of hospital stay per admission			
Less than one week	10	4	2
One to four weeks	15	11	0
Over one month to four months	3	2	1
Over four months	1	0	0
Reasons offered for hospitalization			
Acute illness	7	5	2
Chronic illness	7	1	1
Normal delivery	6	2	1
Obs.-gyn. difficulty, including miscarriages, abortions, and Caesarians	4	1	0
Appendectomy—tonsillectomy	3	4	0
Accident or injury	2	4	0
Mental illness	1	0	0
Average annual incidence rate for hospital admissions per 1,000 population	75	69	37

qualifies many residents of Panorio and Saracatzani, as well as perhaps half the people in Dhadhi, for free care. But even with indigency papers —which take time, effort, formalities, and patience to obtain—the villager must pay $3.30 for admission to the hospital, and he must also pay for his medication. There may also be a wait before a bed in the third-class facilities is available. These facilities, which necessarily emphasize class differences and provide fewer amenities—there are severe restrictions on visiting hours*—distress the villagers, who are sensitive about their status vis-à-vis city people.

The following case illustrates the difficulties the villager often encounters with the hospitals: One of the families of Panorio had already lost several children, and another of their children was mentally retarded. The two remaining normal children were much beloved. One of them, an adolescent girl, became seriously ill, and was admitted after

* Compare city hospital practices, which severely restrict visiting, with those in the country, in which the family moves into the hospital room with the patient (26).

some delay to a hospital as a third-class free patient. Medication and incidentals during the first two months had cost the family $30, which was, they said, nearly the total of their cash reserves.

The mother was agitated about visiting her daughter, for she was unfamiliar with the big city and feared she would get lost if she made the trip alone. Ordinarily she would have gone with her husband, but the cost of their going together ($1.50) was too great a burden. In addition, it was harvest time; if the parents went together, the all-important harvesting would be delayed.

When they did go in together one morning, they arrived at the hospital only to learn that visiting hours were at 4 P.M.* First-class patients could have visitors at any time, but not third-class. The surly—so it appeared—gate guard made them wait for seven or eight hours until it was precisely four o'clock, even though they explained that their daughter was very ill. When they did go in at four o'clock, they found no doctor there, for the doctor responsible for their child worked only in the mornings. No other doctor would see them.† During the three months of their daughter's stay in the hospital—which included two operations—the parents were never able to see a doctor. Finally they had to ask an Athenian relative to talk to the doctor in the mornings. Very rarely was either parent able to see the child; they averaged no more than one visit every week or two. During a crisis period, the mother resigned herself to the fact that she could not be with her child if the girl were to die.

This case illustrates the problems as the peasant family sees them. They were grateful to the government for providing free care, were overjoyed at the eventual cure of their daughter by the competent hospital doctors, and recognized that her life was more important than their own inconvenience, expenses, or sensitivity to slights and low status. Nevertheless, these difficulties made them suffer more than they might have during their daughter's sickness.

The existence of various barriers to hospitalization is used by the villagers as an excuse not to go to Athens and not to seek hospitalization. As long as nature is kind and the sick person recovers anyway, they feel that their procrastination is justified. On the other hand, if the illness

* See Friedl (26).

† We presume that their shyness and lack of information on hospital procedure contributed to the problem, although it is also possible that the other physicians had no time or inclination to trouble with the parents of patients not their own.

gets worse, or if the patient dies, the family is haunted by the consequences of their failure to act. Under these circumstances the natural emotions of grief and despair well up; but because their philotimo requires that they be blameless, they project the responsibility for the disaster outside the family. They blame the government, the doctors, or any other available target. One gentle old man made this comment: "When someone dies [from a curable illness], we have no one to blame but ourselves. We can't really say that we are not without the means to send them to the hospital, but we are so used to saying when someone is sick, 'It will pass, don't bother,' that perhaps we give them a few aspirins for it. But then, if they die, we blame everyone else for it. That is wrong. We are to blame."

HEALING EFFORTS

Families were asked what they had done to cure the illnesses they reported. The most common healing effort during the 12 months preceding the interview was the use of "home remedies," or practika. Eighty-seven per cent of all families said they used one or another form of practika. In order of reported frequency, the remedies used were as follows:

Camomile tea; tea made of other spices (mint, marjoram, anise, sage, and oregano); rubbing and cupping (for rubbing, olive oil, kerosene, alcohol, camphor, herbs, and ammonia are used); holy water (may be drunk, rubbed on skin, or sprinkled in room); aspirin.

Herb teas other than camomile and spice ("white herbs," agriada, polykombo, tea of the mountains, aformoharitis, tilio, kokovaki, sambaroumba, and elafrokeretis); ouzo, either applied externally or drunk; incense, burned during the "xemetrima"; grain teas, of barley, corn tassels, corn coffee, and pomegranate; gum arabic; sea baths and sand baths.

Quinine (without prescription); oil of mouse (an emulsion made by placing a baby mouse in a bottle of olive oil and hanging the bottle in the sun for one year, applied externally); flower teas (carnation); salt, taken as solution; penicillin (without prescription); vinegar (applied externally); bee stings (for rheumatism); hedgehog (ground and eaten); fish intestines (ground and eaten); healing waters, including cathartic spring water from a shrine.

In order of frequency of mention, the following substances were said to be used as antiseptics (use of these is reported without reference to preceding 12 months):

Pharmaceutical antiseptics (without prescription); pharmaceutical antibiotics (without prescription); olive oil; tobacco; salt; soot; oil of mouse; prescribed drugs (after visit to medical doctor).

Soft down of swamp bird's breast; mud or earth (if earth is used, it is to be placed on the wound along with the object that cut or injured the person); stone powder; "balsamo," an ointment; urine (injured person urinates on the wound); porcupine gall bladder; onion.

In addition, the "xemetrima" may be said.

The following foods and drinks were said to be used as external medications (in order of frequency of mention): Olive oil; ouzo; camomile tea; tzipouro (a strong liquor, distilled from grapes); wine; salt; pepper; onion; egg or egg white; flour (paste); cognac.

The following foods are considered to have special healing properties and are given to the sick (some are for specific conditions; others are considered of general value): soups, fish, chicken, meats, yogurt, rice, honey, butter, eggs, cakes, spleen, liver, macaroni, greens, fruits, potatoes, and horsebeans.

The following drinks were said to have special healing properties and are given to the ill (again, some are specific and some are generally useful): milk, camomile or other herb teas, ouzo, dry wine, cognac, sweet red wine, fruit juices, beer, cathartic waters from sacred springs, holy water, coffee, soda, and cocoa.

For sore or infected eyes, including trachomatous eyes, the following unprescribed remedies were said to be useful if applied externally: mother's milk, camomile, collyrium, honey, zamboukos, urine, lemon, grains, egg white, and the xemetrima. For diarrhea, rice preparations were recommended, along with various teas, lemon, and coffee.

In reviewing these data and the substances discussed informally or observed by us to be used as practika, we find the following categories: *

Water	Animal products (excreta, feathers, leather)
Food and salt	Human products (urine, blood, and hair)
Alcohol	Clothing
Plants	Fire or its products (soot and coals)
Stones	Earth
Images and ikons	
Smells	
Ropes, strings, knots	

* It is interesting to note that each of these substances had a magical or healing function in ancient Greece (16, 47, 48, 58, 63). Their continued use today indicates the importance of the ancient heritage in modern folk practice.

The villagers apply their dictum of "nothing in excess" to their use of home remedies, especially food and drink. Too much herb tea will weaken the patient, they say, and the overuse of aspirin is very bad. Even too much water can make a person ill. Implicit in these comments are three themes. One is the traditional belief that the proper life is a life of moderation. A second is that the powers in healing substances might get out of hand; therefore, substances with magical qualities, as many of these are thought to have (wine, olive oil, incense, salt, and soot), must be handled with great care. For the average family or folk-healer there are limitations on the intensity of the powers he can use for healing; caution is the rule. A third, very practical implication of the dictum of moderation is that many of the healing agents are demonstrably potent and produce side effects or even damage if the dosage is too strong. Alcohol and aspirin are examples. The villagers are sensible and observant people; they usually know the limits of their "materia medica."

USE OF MEDICAL CARE DURING THE PRECEDING TWELVE MONTHS

The majority of families in each community (87 per cent in Dhadhi, 88 per cent among the Saracatzani, and 89 per cent in Panorio) recalled that one or more members had gone to a healer during the preceding 12 months. This figure is, if anything, low, both because recall over a long period of time is difficult and because there was evidence that visits to folkhealers—wise women, komboiannites* and sorcerers—were likely to be forgotten or concealed, whereas visits to medical doctors were more quickly remembered.

If we consider only formal visits, defined as those that involve, first, a definite identification of an illness, second, a definite decision to seek advice from someone about its treatment, and third, a formal undertaking to visit or call in a healing specialist for the primary purpose of securing help, we find that the medical doctor is the specialist said to be most frequently used by the people in the three communities. Second most frequently visited is the folkhealer, which here includes the wise woman, gypsy, komboiannitis, witch, or sorcerer. Formal visits to the physicians are reported more than twice as often as visits to the folk-healers in the three communities. Visits to or from a priest or nun rank

* Specialists in hand practika.

third in frequency (these are quite infrequent—they are reported by no more than two families in any one community). Visits to the dentist or the druggist for healing care are least often reported.

The use of folkhealers has several implications. Whereas the frequent use of medical doctors indicates respect for their skills and a level of awareness in the population of the value of medical care, the frequent use of folkhealers indicates a continuation of patterns of the local healing culture, patterns that are at least as old as the written history of Greece. When we consider that from one-third to one-half of the village families have used folkhealers during the preceding year—a usage occurring most frequently among the Saracatzani—we see the vitality of old traditions and the strength of informal within-community approaches to the cure of disease. The infrequent use of the priest in his healing role is discussed at greater length in Chapter 13.

Infrequent visits to the dentist may reflect the relatively low priority given to dental health by villagers. There are no local dentists; dental care is expensive and difficult to obtain, and it is probable that information about dental health has not yet been absorbed by—or even taught to—the villagers. Among the older villagers toothlessness is common, and even among young adults one can observe gaps where lost teeth have not been replaced.

The villagers often use the various treatments they describe simultaneously or in close sequence. The frequent use of home remedies does not rule out visits to a doctor, just as going to a doctor by no means rules out the simultaneous use of a folkhealer or a priest. Generally the progression is from the easy to the difficult, from the family to the non-family healers, from the near-at-hand to the far-away, and from the free to the costly.

Efforts at treatment are typically empirical, having a trial-and-error pattern: if one method fails another is tried. As long as expense and effort are not involved, the villagers are completely open to experiment. "Why not? It can't do any harm" is the way they usually express their attitude.* Even though the responsible physician may have made a diagnosis, there is no assurance that the patient will believe what the doctor has told him. On the contrary, the patient and his family may visit two

* The philosophy of "it can't do any harm" applies only to the use of natural substances such as herb teas, mouse oil, balsam, etc. Manufactured pharmaceuticals are viewed with suspicion, as will be explained in Chapter 7.

or three other physicians just to see what they have to say. Even when the physicians agree on a diagnosis, the family diagnosis of the evil eye may still be credited and magical treatments continued along with visits to the physician. Even in the hospital, folk medications may be used.

The peasant sees no contradiction in his simultaneous use of treatments based on different assumptions about illness. This is not inconsistent to him because he believes that diseases have various causes, each of which has its own forms of cure. The niceties of city logic do not bother him; he merely seeks a cure by any means available. Implicit in his attitude is the notion that the wise man "plays it safe." In a culture where there are several independent systems available for explaining the causes of illnesses, one does well, in seeking a cure, to cover one's bets. The Greeks are marvelously open-minded; their belief systems are not closed or dogmatic. They believe one must always consider alternative possibilities. It was this flexibility of spirit, perhaps, that led the ancients to erect a temple to the Unknown God, and that inspired the current story of the man who was found lighting two candles in church, one to God and the other to Satan. When asked what he was doing, he explained, "Okay [to light only one candle] if I go to paradise, but what if I go to hell? I'd better be on good terms with the devil as well."

ORIENTATION TO TREATMENT ACTIVITIES

Earlier in this chapter barriers to the peasants' use of hospitals were cited. We shall now comment further on factors that influence the speed with which various treatment activities are undertaken.

Health is greatly valued in Greece, just as it was when Gorgias said, "The highest blessing possible for a man to possess is the health of the body" (52). The common toasts nowadays are wishes for good health—"Health to you"; the common greetings are "May you go well," and "May your children live for you to enjoy them." Among the conscious goals in living are to remain as healthy as possible and, at death, to give up the soul easily without having to suffer a terminal illness. The church affirms the hygienic value, saying that the body is the temporary shelter of the soul and as such must be protected and cared for to give the soul a good earthly home. Parents naturally want their children to be healthy and above that, to have excellence, the source of parental pride. It is hoped that a child will excel not only in his conduct and filial loyalty, and in his studies, work, and marriage, but also in his body, so that he will have strength, agility, beauty, wholeness, good features, strong eyes, and a good complexion. Parents hope that their boys will be tall, muscular, and mustached, and that their girls will achieve the plumpness necessary for beauty and for bringing a well-nourished infant into the world. They realize, however, that these goals are not easily achieved or maintained. Their proverb says, "Health leaves by the sack and returns through the needle's eye."

The value placed on health and the awareness of the ease with which it can be lost are the fundamental reasons the peasant seeks treatment. He is quite aware of the importance of health, but, unfortunately, he does next to nothing about preventive measures or treatment. The reasons are varied. For one, the circumstances of living make it difficult to take time out for activities that do not seem to be immediately necessary for the maintenance of life. A man is unwilling to leave his work in the fields to attend to a stomachache; a woman is reluctant to abandon the incessant, pressing demands of house and fields to get a diagnosis for seemingly unimportant inter-menstrual bleeding. The villagers are also reluctant to take steps for illnesses that they know need treatment, but for which they expect treatment to be expensive. In any event, any treatment other than home remedies disturbs the daily routine, threatens loss of time from urgent work, and may cost money, which is at best spent grudgingly on health. If the illness appears to be mild, or does not threaten their livelihood or their family and community position, or if it is chronic, they may be expected to delay seeking anything but local remedies.

The patient's own diagnosis of his illness is a second major factor in determining whether and from whom he will seek treatment. Because the peasant believes that a head cold is not a sickness, or that a pain in the back is one of those conditions "doctors do not know about," he will not consult a physician about it. The definition system applied will shape the steps taken; in the case of a backache, the "waist" will be brought to the attention of a hand practikos, a healer, or a wise woman. Ignorance, as well as folk concepts, plays an important role here; a person who does not know that shortness of breath and edematous ankles may be symptoms of cardiac trouble has no reason to think himself ill and to initiate diagnostic or treatment measures.

A third set of factors are physical ones that we have encountered before: distance, inconvenience, lack of local facilities, cost, and, for women especially, lack of independence—they may not be able to leave the house for a trip to town without the aid of their husbands.

We have seen that socio-psychological factors are important in determining hospital use. Peasants are sensitive to discrimination by physicians and hospital personnel, who they believe slight the country folk who are poor, unsophisticated, and lacking in influence (messa) with important people. In addition, the villagers are fearful of what the doctors will find wrong with them, are anxious about going into the strange and confusing big city, and are sometimes dissatisfied with the medical care they have received in the past. Out of fear, some may deny pain or disability, or may put off going to a physician if they suspect an illness that is fatal. Anticipation of the worst and consequent delay can occur especially in the case of ambiguous signs or symptoms that do not conform to known ordinary illnesses. Shame may also prevent a person from seeking treatment; if he suspects he has a disease like tuberculosis, venereal disease, epilepsy, or a psychosis, he may be afraid of exposure and the associated threat to family honor and personal philotimo.

Other factors also operate. The intensity of pain and the extent of disability no doubt influence treatment priority. The relationship of the villager or his family with various healers will determine the choice of a healer. It also appears likely that the choice of treatment modes is related to (1) the amount of formal education the villager has and the ideas and status he associates with being modern; (2) the responsibilities of his family role, in which attending to work or to others takes priority over seeking care for himself; and (3) the attitudes of his family toward his illness, attitudes that either encourage or discourage him in seeking medical attention.

INNOVATION

The trial-and-error approach to sequential and simultaneous treatment offers the villagers the chance to try many kinds of innovation. If someone they trust and admire uses a new item, if a new method can be shown to have good results without entailing any major changes in life style or social practices, or if a suggested device looks as if it might work because it is similar to ideas or tools already possessed by the peasant, the villagers readily accept new things. Acceptance of innovations increases as cost declines; and in Dhadhi, at least, acceptance seems to be influenced by the fact that use of the new remedy lends the user prestige for being modern or citified. The villagers quickly recognize the value of new drugs such as the antibiotics and penicillin, and they even prescribe them for one another, offering whatever pills they have on hand. They reason that "if it is good for one thing it must be good for another."

Children play an important role in the adoption and spreading of new health ideas. They are eager for learning and enthusiastically take on the job of teaching their parents. During our study, a Greek National Red Cross mobile unit came through Dhadhi and left pamphlets at the tiny store to instruct people in health protection and specific preventive measures. The children took the pamphlets, chattered about their contents, and tried to communicate their interest to others. In many instances the adults told us that they had learned the answers to the hygiene questions in our interview schedule from their children, who had been taught these matters in school.

Wine and ouzo have for millennia been considered therapeutic and are taken internally and applied externally. In accordance with this practice, cognac (bottled Greek brandy) is now added to the nostrums; it is made from the familiar grapes, consists of trusted alcohol, and has the added attraction of being a glamorous city product that is powerful and therefore "must be good."

One day when we were in Dhadhi, an energetic young man showed us a brass bracelet for which he said he had paid $10 (no doubt an exaggeration but a near fortune at half the price). He was delighted with it and said, "Now you can throw away all you've written about that practika. Now we have science in our village. Look, here is the twentieth century on my arm!" He explained that the bracelet had been made in Japan, that it was magnetic, and that it would work on the blood circulation, the nerves, and the humors or hormones. Since these three elements were "the basis for the body," the bracelet would cure may things, including high blood pressure, kidney trouble, rheumatism, and impotence.

He told us that he had learned about these recently imported items from his koumbaros (his godfather who is therefore "adopted" kin and one to be trusted),* and having tried it on himself this day, he intended to sell it to the other villagers. We challenged him about its high price, but he said it was no more than one would pay for a visit to the doctor

* The word also means best man at a wedding.

(from which we gathered that he either had purchased it for one or two dollars, which is the doctor-visit fee, or was practicing his sales spiel on us), implying that it would cure a great deal more than the doctor would.

We do not know whether he sold his neighbors on the product, but the incident illustrates how a familiar principle can be accepted in a new form. Magnetism is a known phenomenon that reputedly has magical power; it is the "magnetism" of the eyes that bewitches in the evil eye. The belief that the bracelet would put "magnetism" to work for one's benefit demonstrates the accepted principle that power is neutral, and that the goal of magic or manipulation is to invoke and direct that power. In this case the bracelet became an amulet, a familiar enough device to the villagers since most of them use charms on their animals and babies to ward off the evil eye. It had the further advantage of being "scientific"; i.e., it embodied a force used in scientific and technological work, and therefore had the prestige of being progressive.

7

Response to Medical Examination

Our purpose in introducing a public-health team into the three communities was threefold. One aim was to provide medical service to alleviate some of the immediate physical distress of the villagers; a second aim was to gather morbidity data, as presented in Chapter 5; and a third aim was to observe the behavior of peasants and shepherds in a medical-care situation. This last aspect provides the content of the present chapter. We shall discuss here the satisfaction of the patients with the medical examinations and their understanding of and cooperation with the medical advice they received.

When we informed the villagers that public-health teams from the Ministry of Health, the Attica Health Center, and the Children's Hospital would be coming to give free examinations, only two families were indifferent. The rest were happy that free care would be available. Many of them spoke of current ills and how they would not have been able to afford a visit to a private physician for care. The gratitude to the Ministry was, in some cases, quite touching, and in marked contrast to the distrust and dislike that sometimes characterize the attitude of rural people toward governmental agencies.

Villagers were prepared for the visit of the medical teams by at least two calls from our team. During these calls appointments were set up so that the limited time of the public-health teams would be utilized fully. Not all of the family members intended to see the doctors, even if they had said they thought the presence of a medical team was a good idea. In Dhadhi 14 persons said they would not come, and several were uncertain. In addition, four families could not be reached or were unable to say what they would do. On the basis of these replies, we estimated that a maximum of 36 persons would not appear for medical examination in Dhadhi. Actually, fifty-nine did not appear. In Panorio, 15 persons indicated they would not or could not come, and two others, plus two families, could not be reached or could not say. Here we estimated that 26 people would not participate. Thirty-four did not appear. Among the Saracatzani everyone said he would try to come. Nevertheless, four did not appear. But, of these, two or three would have come if the medical team had not been delayed on its scheduled day; the shepherds could not wait and had to go out to work.

We found that the most frequent excuse given by those who did not intend to come was that men working as laborers for others could not take time off, or that women were too busy in the fields. The statements that they "did not need a doctor" or that "nothing is wrong with me" were also common. Other excuses included the belief that going to a doctor was a "bad omen," and that the doctor would give one "the idea" of sickness and thereby make one ill. Another frequent excuse was to the effect that if one's good health were affirmed by medical investigation, it would become known and might make one vulnerable to the envy and admiration of others. One would thus become vulnerable to disease through the destructive wishes of men or the gods. This notion has elements in it of the old "hybris" belief that the gods strike down those who are prideful or upon whom fortune visibly smiles. The gods, of course, can act through the malice of one's neighbors. The idea here is similar to the idea of "jealous diseases" (as discussed in Chapter 9), in which a sick person, or, more precisely, the spirit of his disease, seeks to infect a healthy person.

Several older women had never been to a doctor, and said they were too shy or fearful to go. The family of one of these women, after family council, agreed to support her in her decision and decided that none of them would attend. (Nevertheless, they did appear because at a second council they decided they did not wish to disappoint us.) Others said they had no confidence in doctors, that doctors could not cure the illnesses they had, that seeing a new doctor might lead to interference in current care by a private physician, or that they would be away from the village while the medical team was there. One woman did not want to take a bath as we requested her to do before coming to the examination (she needed one badly), and one woman said bitterly, "We need food, not doctors." Several children were too afraid to come.

The fear expressed by several women that others might learn about their ailments and that they would thereby be ashamed or made the butt of gossip proved to be an avoidable obstacle to participation. We assured them that separate examination rooms would be set up in the schoolhouse, that no one would be present during the examination except those they wanted there, and that the physicians would be instructed to speak quietly and under no circumstances to divulge information to other villagers about anyone's condition. We fulfilled our promises by dividing the schoolroom into cubicles with sheets, having villagers wait outside the schoolhouse, and reminding physicians to keep their voices low. By learning about these fears in advance and by taking visible steps to protect the interests of the patients, we were able to enlist their participation.

PARTICIPATION IN THE MEDICAL EXAMINATION

Between three days and two weeks after the medical examinations were completed, each family was again interviewed. Those who had not come in for examination were asked why they had failed to appear. The two reasons most commonly given were that they could not take time out from work or that they were "healthy," but, paradoxically, had been afraid of what the physicians might find. It became apparent that many who did participate had been hesitant about it and felt some of the same shyness and fear, presumably to a lesser degree, that had been voiced before the examinations by their neighbors who did not come for examination. In spite of considerable preparation and persuasion on our part, which included drawing on the reserves of goodwill, trust, and obligation that our team had stored up, the villagers' pleasure at the prospect of seeing a doctor had often given way to reluctance. However, Table 11, which presents information on the proportion of villagers participating in each phase of the examination, shows that nearly three-fourths of all the villagers did participate.

The Saracatzani, with whom our team had the warmest friendship, provided the greatest proportion of cooperating persons. But cooperation in all the villages could be considered good. Over-all participation of more than 70 per cent in the three communities contrasts with the estimate of 33 per cent voluntary rural participation in the most recent mass chest x-ray surveys by a mobile unit (35). This contrast suggests that preliminary preparation through intensive field work and establishment of ties of trust and affection with the villagers can make a considerable difference in the extent to which they will participate in medical care.

AGE AND SEX CHARACTERISTICS

Table 12 shows that children were best represented in the medical examination. Mothers and fathers made a considerable effort to see that their children saw the doctor. The group least likely to come for exami-

TABLE 11

Cooperation in Medical Examination and Laboratory Studies by Residents of Three Communities

Category	Dhadhi (N = 203)	Panorio (N = 125)	Saracatzani (N = 41)	Total Per Cent, 3 Villages
Volunteered for exam. .	144/208 (69)	97/127 (72)	37/41 (90)	73
Gave blood for analysis[a]	124/144 (86)	86/91 (95)	32/37 (87)	82
Gave urine for analysis[a]	104/144 (72)	85/91 (93)	29/37 (73)	80

Note: Changes in population (N) for each community reflect marriages, births, and visiting relatives added to households as the study proceeded.

[a] Numbers in parentheses are percentages of the number examined who participated in the laboratory studies.

TABLE 12

Characteristics of Nonparticipants in Medical Examination for All Three Communities

Age and Sex	Did Not Have Exam	Did Not Give Blood[a]	Did Not Give Urine[a]
Boys 0–16	5/49 (10)	9/44 (20)	13/44 (30)
Men 17–69	62/114 (54)	4/52 (8)	4/52 (8)
Men 70 and over.	3/15 (20)	0/12 (0)	0/12 (0)
Girls 0–16	7/67 (10)	9/60 (15)	18/60 (30)
Women 17–69	18/12 (16)	8/94 (9)	19/94 (20)
Women 70 and over.	3/14 (21)	0/11 (0)	1/11 (9)

[a] Numbers in parentheses are percentages of the number examined who did not participate in the laboratory studies.

nation was that composed of working-age men. Over half of the total population of working men failed to come for examination. The reasons cited earlier—unwillingness to lose work time and the paradoxical attitude, "I am healthy, but I fear what the doctor might find"—were most often given by this group. It is apparent that a father may insist that his children and wife see the doctor but will himself decline the opportunity.

About four out of every five who came for examination were willing to give blood and urine for analysis. This high degree of cooperativeness contrasts with the belief frequently expressed by Greek physicians that peasants are unwilling to give blood. Although it is true that many were reluctant about it, the presence of others, friends and family, and the friendly persuasion of the laboratory team (who handed out balloons to the children) helped to overcome resistance.

It can be seen from the age-sex breakdowns in Table 12 that males and children were the two least cooperative groups in the blood examination. Male children ranked first in refusals. Females of all ages were much less afraid of having their blood taken than many of the men, who cooperated uneasily. Children were more often uneasy than the aged.

Children were least likely to give urine, partly because some were unable to urinate at the time of the examination and partly because the follow-up visits often failed to catch them at a favorable moment. Among the women noncooperation was more often a matter of shyness or shame, especially if they were menstruating at the time, for to give urine with blood in it would be to display their state of pollution.

ROLE IN THE COMMUNITY

Another factor affecting cooperation in medical care was the social position or social role of the individual. For example, in Panorio there were four people who were "deviants," i.e., unmarried adults. These persons considered themselves somewhat isolated from village life and especially subject to criticism; their self-appraisals were substantiated by the opinions of others. Those who did not marry were less esteemed, more rejected, and less often part of the family communication nets of the village. Three of these four single people, one spinster and two bachelors, failed to come for medical examination.

In Dhadhi there was only one peripheral person without any family ties, a hired hand who lived with a family. In another family there was an "adopted soul son," an older laborer who had lived with the family for many years. Neither of these men appeared for examination. This was most likely due to the fact that they were working-age men, but the fact that they were "deviant" and somewhat isolated from the community probably played a part in their noncooperation. They were less subject to family pressures and were not likely to participate in the family discussions that helped prepare many of the villagers for the anxiety-producing experience of coming to the doctor.

In Dhadhi there were three lechones (women who had given birth within the last 40 days) at the time of the medical examinations. Each had been reluctant to break the taboo that forbade her to leave the house during the 40-day postpartum period. Nevertheless, these young mothers were convinced of the value of a physician's seeing their babies, and therefore did come for examination. They rationalized their fears of what the village gossips might say about their nonconformity and of the unknown magical dangers that threatened their babies and themselves by saying that the presence of the physicians was a special circumstance that would be free of threats from magic or pollution. They were also aware that in the medical setting the taboos against their own presence as polluted persons dangerous to others would not be operating. In their fears, conduct, and rationalizations we have a good example of how a person can compartmentalize beliefs so that ordinary taboos do not come into play. Indeed, the whole village joined these mothers in this, separating the medical activity in the schoolhouse from thoughts of taboo and pollution. The villagers accepted the "city" system of rational medical science and acted fairly consistently within its assumptions as far as the lechones were concerned.

On the other hand, two of the three lechones did not give urine samples. This may have been because they were very busy with and concerned about their babies who were being examined (one baby was quite ill). Their failure to give a specimen may also reflect the same concerns of menstruating women. The taboo against blood seems to be much more powerful than the taboo that isolates the lechone. Mother-love encourages a woman to bring her baby to the physician, but the taboo against blood discourages her from giving a urine specimen.

In contrast to those villagers who failed to participate because of their social positions—the isolated unmarried adults and the socially and magically vulnerable lechones—those villagers who held positions of leadership invariably participated in the medical program. In Dhadhi, it was in the school and under the auspices of the teacher, the community leader, that we were introduced to the village; and it was in the school some five months later that the medical examinations were held. Both at the beginning and at the end of the study the teacher exerted his authority to order the villagers to participate. As a high-status person in the village, he had immediately considered himself as having an equal-status link with our team. Because of that image—in his own if not in the villagers' eyes—he was consequently obliged to participate in the medical examinations. It should be noted, however, that he participated in spite of his evident reluctance, which presumably stemmed from his fear of what the doctors might find, and perhaps from his dislike of finding himself classified on a level with the other villagers as one asked to participate. The teacher's family was also under pressure to cooperate in the examination. But since his wife had a special healing role in the village—she was the one who gave first aid and who owned the only first-aid kit within many miles—her participation was compatible with her interests in public health. She and her husband collected materials from the villagers and helped prepare the schoolhouse as the examination center.

The two progressive elders of Panorio expressed their support of the medical examinations and cooperated in the program. Neither made any formal effort to persuade others to come, but we may assume that their persuasive efforts in quiet conversations with others had some effect. Indeed, such efforts were necessary within their own families, for the wife of one of the elders feared the medical examination because she had never before been to a doctor. Her fears brought her into a psychological conflict. Nevertheless, the elder's image of himself and his family as progressive, his demand on them all to live according to reason, and, in addition, his family's affectionate ties with our team, finally resolved the conflict in favor of participation with us.

Our first contact with both Saracatzani family camps was through the eldest brothers, the undisputed leaders of the two encampments. They were warm, intelligent, and humorous men who quickly saw to it that the entire camp cooperated in all phases of the study, including the medical examinations. Communication within these family groups was immediate and authority was clear-cut; since all individuals were operating within a family framework of trust and mutual obligation, there were no apparent efforts to rebel or undercut authority, nor was there any opportunity to evade participation, which was undertaken as a clan enterprise. Here the leaders themselves cooperated fully, and saw to it that everyone in the family did likewise. As we indicated earlier, Saracatzani attendance at the medical examinations would have been 100 per cent had it not been for an administrative mix-up in Athens, which, in depriving the health team of its transportation, forced a delay that the work requirements of the shepherds could not tolerate.

MEDICAL EXAMINATION FOLLOW-UP

Another purpose of our interview with each family after the medical examinations was to determine the extent to which the family had understood what the doctor had said, their degree of satisfaction with the care they had received, and the extent to which they cooperated in obtaining recommended prescriptions, in taking drugs given them, and in seeing specialists.

To measure their understanding, we compared the physician's entry on the examination form under the section, "State exactly what examining physician has told or recommended to the examinee," with the patient's response to the interviewer's question, "What did the doctor tell you?" This comparison revealed that nearly every patient did understand the substance of the doctor's remarks. In Dhadhi, 94 per cent had understood; in Panorio, 82 per cent had understood; and among the Saracatzani, 79 per cent had understood. Although the majority did understand, we should not discount the importance of the fact that one of every five persons in Panorio and among the shepherds was unable to repeat anything the doctor had told him.

If we compare the villages, ranking them in order of the proportion of the population who understood the physicians' statements with the relative education of their inhabitants, we see a perfect correspondence. The average education (number of years in school) was highest in Dhadhi, followed by Panorio and the Saracatzani. As we saw, Dhadhi had the largest proportion of patients who understood what the doctor said, Panorio ranked second, and the Saracatzani ranked third. It does not seem unreasonable to suggest a relationship between the adequacy of physician-patient communication and the educational level of patients.

In the follow-up interview each family was asked what it had thought of the examination, and whether it was pleased or displeased with the doctor and his work and personal approach. The majority were pleased;

79 per cent of those replying in Dhadhi said they had no complaints; 65 per cent in Panorio and 90 per cent of the Saracatzani were also pleased. The dissatisfactions that were expressed took three major forms. One was that the examination was too quickly and superficially done; the second was that the physicians appeared uninterested or rude; and the third was that not enough free drugs or special procedures (x-ray, other specialists) had been provided. It was a matter of concern to us that nearly one-third of the villagers in Panorio were dissatisfied. We suspect that several factors were operating to produce this dissatisfaction. One is that unreasonably high expectations had been generated—expectations for continuing protective care and full payment by the team for all needed drugs and medical services. These expectations, sometimes stated explicitly, seemed to reflect intense dissatisfaction among the Panorio villagers with their present poverty. This dissatisfaction, coupled with a psychologically dependent attitude toward outsiders, whether government or "these rich Americans," led them to demand that the "powerful" outsiders relieve them of their burdens. It was understandable that they should express these feelings in the form of high hopes, hopes that could not be met by the limited "one shot" approach to care, which was the only one possible with the resources of the study.

Dissatisfaction with medical care based on the failure of others to meet unrealistic hopes for cure or help or love is also found among patients in the United States, where studies (9, 11) have identified some of the personality dynamics of people whom American physicians are correct in describing as demanding, unreasonable, and unrealistic. Generally, such patents are immature. They depend heavily on others for their needs, expecting them to perform tasks that they are unable to do for themselves, asking for miracles, which, at one level of thinking, they realize cannot be achieved. Expecting others to do things for them, they often fail to cooperate in medical care. When something goes wrong, or when their high hopes are not met, these unrealistic patients blame the physician; they cannot accept responsibility or blame for themselves.

Unreasonableness, unrealistic expectations, and unwillingness to assume responsibility seem to exist, as a personal rather than a cultural variable, among rural Greeks just as among American city dwellers. Here are two illustrative cases:

One woman brought her sick child to the public-health doctor. He diagnosed bronchitis, and told her to put the child to bed and to buy some cough-syrup–expectorant that he prescribed. For several days after this we saw the child running about, obviously not having been put to bed. Three days later the woman stopped us, complaining angrily that the child still had a fever. She complained further that the doctor had not given her child the vitamins that he had given to the other children (a limited supply was distributed, without cost, to undernourished children). She said she wanted to see a doctor again. We said it would be a good idea. She replied that she would do so but would refuse to see the doctor she had visited before because he had not cured the child and was therefore not a good doctor. Inquiry indicated that she had not purchased the prescribed medication.

One Saracatzani woman from a neighboring encampment not included in our study insisted that she be allowed to come for an examination. Our policy was not to refuse outsiders such care, but not to encourage it because of our very limited time and resources. We inquired about her health, and learned that she had had a complete hospital examination just 20 days before and had "been going to doctors for years without their finding anything wrong." We told her it was unlikely that our doctors would do anything better in view of their limited facilities. (We were careful to conduct our discussions with her around the Saracatzani campfire for all the shepherds to hear, for her angry manner warned us that she would be dissatisfied afterwards and would be a gossipy blame-thrower.) She came for the examination and afterwards reproached us bitterly. Our doctors had found that she had high blood pressure and exophthalmic goiter. She was furious that they had found something wrong with her after she had gone to the hospital and "they found nothing wrong!" She preferred a different diagnosis.

During the next few days she approached us on several occasions, berating us about our doctors and then demanding of the physicians with us (not members of the public-health team) that they discuss her case. Among other things, she said she wanted to go swimming; now that she had high blood pressure she did not know whether she could bathe in the sea. She was obviously very anxious about her health but was expressing her anxiety in an angry manner. She was also exaggerating the common Greek procedure of seeing several doctors by seeing many and then being angry with each for not reducing her psychological distress. No doubt she would benefit from psychotherapy, but no physician made such a referral, nor could he do so, since facilities for out-patient psychotherapy do not exist in Greece. This woman is a clear example of the unreasonable patient who does not cooperate and who is dissatisfied.

Another discernible factor that produced patient dissatisfaction was a more legitimate one. Some of the complaints about the quality of the examinations and the attitudes of the doctors were consistent with impressions gained by outside observers. However, one must keep in mind that the public-health doctors were forced to work under great pressure, and that their working conditions were less than ideal.

RESPONSE TO PHYSICIAN'S RECOMMENDATIONS

In reviewing the reports of medical examinations as completed by physicians, we find that a number of specific recommendations were given to the villagers. Three out of four patients who saw the doctor were given some treatment advice. This is an important figure in itself, for it suggests the proportion, remarkably constant from one community to the next, of villagers who may be expected to receive potential benefit from medical attention. The questions arise, however, whether many of those who were offered treatment advice did in fact act on it, and whether many others disregarded it or were unable to follow through.

Within a period of three days to two weeks after the medical examinations, the interview team made an effort to see especially those patients who had received some form of instruction from the visiting public-health physicians. It was not possible to see everyone in each community during the time available, nor was it always possible to determine what action they had or had not taken in response to the physician's instructions. But insofar as it was possible to contact individuals and get their

TABLE 13

Reported Cooperativeness Among Patients in Response to Medical Recommendations

Category	Dhadhi (N = 52)	Panorio (N = 29)	Saracatzani (N = 22)
Per cent following all of the recommendations...	51	35	77
Per cent following only part of the recommendations	2	20	5
Per cent following none of the recommendations.	47	45	18

Note: Data gathered at first follow-up visit, made within two weeks after the medical examination.

TABLE 14

Reported Cooperativeness in Response to Recommendation to Visit Athenian Specialists or Laboratories for Follow-Up Examination

Category	Dhadhi (N = 31)	Panorio (N = 32)	Saracatzani (N = 7)
Per cent reporting visit to Athens	19	19	57
Per cent reporting no visit to Athens but expressing intent to go	29	22	0
Per cent reporting no visit to Athens and expressing no intent to go	49	59	43

Note: Data gathered at second follow-up visit, made within four to eight weeks after the medical examination.

reports, we found many who had not cooperated with medical care, if by cooperation we mean doing what the physician had recommended. Table 13 presents the available data on the cooperativeness of those receiving advice in the three communities.

The second follow-up, referred to in Table 14, was made as a special effort to see a subsample of individuals on a second occasion, timed somewhere between four and eight weeks after the health-team examinations. In the second follow-up, persons who had been told to see specialists in Athens for diagnostic tests or treatment were asked if they had done so. Arrangements had been made, through the Ministry of Health and the Attica Health Center, for all such visits from patients in our study to be free and facilitated by the public hospitals and clinics. As it turned out, these arrangements sometimes broke down.

Table 13 shows that the Saracatzani were the most cooperative group in responding to medical recommendations for drugs and regimes. A few people volunteered explanations for their lack of cooperation. Some said they did not like the medicine or could not afford it; others said they had taken it for a time, but that it had not helped them. Practical reasons stood in the way of many; during a season of heavy work, sea baths, which were recommended, were not practicable, especially for Panorio villagers, whose village was a long, hot, mountainous walk from the sea. One can imagine that those who had been told to eat more food, or to have a more nutritious diet, would have been happy to comply. But even if they had had the money to buy meat or fish, the lack of refrigeration for keeping it would have made it impossible for them to follow the doctor's advice.

It is understandable that reactions were negative to diet restrictions on the few foods in common use—bread, oil, vegetables in season, cheese, and wine. Eating in Greece is charged with significance above and beyond subsistence or epicureanism. As we noted in Chapter 2, giving food is a way of loving and strengthening; eating is to accept hospitality and love, and to feel the increase in strength against unknown future dangers. Any dietary advice, then, intrudes on individual and family life and the deep-seated beliefs about personal power, health, and beauty, which are secured through food.

Advice to take better care of oneself, to rest more, to sleep more, to wash more, makes heavy demands on an already overburdened schedule. The 16- or 17-hour workday of some poor peasant families and the rigorous challenge of the rocky soil and scarce water make it understandable that they cannot "take it easy" merely on a doctor's advice. Similarly, when the water is far away, must be carried by donkey or on one's own back, has to be used sparingly for drink, and must be heated over a twig fire, the advice to be clean seems like too much trouble, especially to those peasants reared without any notions of hygiene or body cleanliness.

DRUGS AND DISTRUST

One important factor in the refusal of prescribed medicines was the distrust of strange or "foreign" substances. This distrust contrasts with the trust in natural or home-grown foods and drinks and in the recommendations made by trusted persons such as family or local folkhealers.* Trust and distrust are key dimensions in any culture; learning who and what are distrusted is a pathway to the understanding of much in the social life and belief of a culture. In Greece the trust-distrust dimension is dramatically imposed on many kinds of social acts. City people, strangers, and government officials are immediately distrusted, whereas the family, and to a lesser extent the immediate community, are trusted (but not necessarily regarded with affection). The peasants trust the familiar—the local world and the natural surroundings of the plain or

* Foster, in his book *Traditional Cultures and the Impact of Technological Change* (22), reports that peasant distrust of outsiders is common throughout the world. He cites a Ceylonese experience in which villagers refused to take pills after hearing the rumor that the pills were miniature delayed-action time bombs that would explode inside them. He relates this distrust to the poor treatment villagers have received in their dealings with cleverer and more powerful city and government people. Although this may have some role, the generalized distrust of others within the same village can also be regarded as suggesting a common psychological substrate for the approach to the strange.

valley close to home. They distrust anything that comes from afar, has strange shapes or smells, or is given by people who are neither kith nor kin.

We have mentioned that food has an emotional and social significance in rural Greece that it does not have in northern Europe or the United States. The mother feeds the baby (overfeeds it, to our way of thinking) so that it can grow fat and thereby have the strength to withstand the inevitable challenge of illness. Giving food is a communication of love, a way of building a relationship, warding off anxiety, doing magic for building power to withstand future evils, and making children, friends, and guests feel wanted, at home, and protected. People are proud of how much they can eat. The foods they offer are natural foods, the known products of local labor or the generosity of the earth. They are part of the structure of solidarity, the sharing of life and emotion, and the unity of trusting interdependence. But foreign objects, strange pills prescribed by strange doctors and made by strangers in factories far away, are substances to be avoided unless the illness is acute.

The extent of this block against the strange is shown in the case of a woman from Panorio, a gentle creature who, during the course of our study, spoke to us of her cough, weakness, and mild but chronic fever. Although the final diagnosis of her illness had not been made by the time we left the area, she had been put under care by the Athens clinic with a provisional diagnosis of tuberculosis. She recognized her illness as dangerous and debilitating, and made many efforts to cooperate. But each time she took her pills she vomited them. She could not reconcile her need for medication with her deep-seated distrust of foreign objects. Such strong beliefs about the dangers which arise from the incorporation of foreign objects may be seen in non-literate societies. In such societies one of the most common healing methods consists of the ritual extraction of "foreign objects" (imagined, magical, or real) believed to have become lodged in the body and to have caused illness (25). We did not encounter this illness explanation or healing method in rural Greece, but we believe that the resistance to taking drugs stems from similar psychological processes.

The fear of medicines is not limited to a distrust of the intent of outsiders and the substances they make and prescribe or to the anticipation of untoward immediate effects. There is also a common belief that medicines will stay in the body so that when one dies, one's body will not decay. The decay of the body is deemed necessary for the release of the soul so that it can journey to the next world. Those whose bodies do not decay are accursed, fated to linger on earth as suffering revenants, or "vrikolakes." Because of this belief, there is reluctance to take medicine for fear of being condemned after death to this dread state. An illustration of this belief is found in the following conversation (in which the grandmother is an old woman thinking about her approaching death):

Grandmother: I wish I could die without suffering.
Woman visitor: I'll give you some medicine to help you die.
Grandmother: No; if I took it, I'd lose my soul, for they say you don't dissolve. You remember what happened to that girl in Doxario? God didn't take her. Not that any of us have gone to heaven to know, but that's what they say happens.

This particular fear of drugs is complicated by the belief that suicides are destined to become vrikolakes, as was the case, so the villagers believe, with the Doxario girl who killed herself with drugs. Not only is death associated in thought with medicine, but the Greek language itself supports the link: the word for drugs or medicines is "pharmaki," which also means "poison."

Returning now to Table 13, we may summarize it by saying that about half of the peasants and three-quarters of the shepherds followed all or part of the doctor's advice within, or for a period of, at least a week or two after the examinations. There are a number of possible explanations for noncooperation, most of which seem to be related to distrust of manufactured drugs and medicines, a lack of understanding of the role of medication in the treatment of their complaint (part of a larger failure to comprehend bodily function, medical science, and therapeutics), and an inferred skepticism about the efficacy of the medical recommendations. Drug cost and the inconvenience of going to a faraway pharmacy were also deterring factors.

GOING TO ATHENS TO SEE SPECIALISTS

Table 14 shows that the Saracatzani were again the most cooperative group in seeking recommended examinations by specialists or laboratories in Athens. Half of the eight shepherds advised to go to Athens did so, whereas only about one-fifth of the peasants in each of the two other communities did. The degree of peasant cooperativeness would increase if statements of intent to follow advice in the future were taken into account. We do not know whether the villagers' stated good intentions were carried out, but we suspect that a good number of them did not go to Athens, and thus belong in the uncooperative group.

The reasons offered by those who did not go to Athens for the recommended care are not new. Some complained of the expense of the round-trip bus fare (approximately 80 cents) plus the loss of a day's wages for laborers working for someone else (between $2 and $3 for a man and somewhat less for a woman). Others returned to the theme that it would be a "bad omen" to go, that further visits to the doctor would only be inviting the "idea" of the illness. A few said, quite frankly, that they were afraid of the bad news the doctor might give them about the state of their health. One old woman said she was too old for surgery and expected to die with the ailment (an umbilical hernia) with which she had lived so long.

From the villagers' behavior, we inferred another reason for their avoidance of the trip to Athens. The big city can be a bewildering place for the rural visitor, and for the unadventurous it is a frightening place. The villagers were easily confused about buses, locations, and procedures. Even in the sheltered circumstance of being examined in their own village schoolhouse, they were disoriented by the novel arrangements. Unless they were given the most careful instruction and sometimes physical guidance, they lost their way between the examining room and the laboratory room; the sequence and activities were totally strange to them. Had they not been literally led by the hand, they would have sat holding their urine bottles for hours, having no idea to whom they should give them. When we went with them to the Athens clinics and hospitals, we saw them become totally disoriented, shy, abashed by their own ignorance, and painfully sensitive to the abruptness or implied disdain of clerks, guards, and minor hospital officials. They were very willing to go to the Athens clinics in our company, for they were confident that we would guide them, and that with our "messa" we would procure for them attention and protection from insult and the ever-present threat of ridicule. But without such a guide, many of them quailed at the thought of the trip.

DOCTOR-PATIENT RELATIONS

Another set of factors that appeared to influence patient cooperation was associated with the kind of relationship that was established between the physician and the patient. We have already mentioned that some of the dissatisfied patients had been annoyed at what they considered to be hasty examinations or disdainful attitudes on the part of the doctor. Obviously these were responses of a minority of patients, for most, as we have seen, were not displeased with the physicians. Direct observation of physicians and patients during several hundred of the examinations indicated that most patients accepted the directness of the physician (termed discourtesy, disdain, or a peremptory manner by some observers) as the natural way for an authority to deal with a villager. But some rebelled because they felt it was a slight to their philotimo or another painful demonstration of the status difference between city and country folk. Consequently, the irritated or rebellious patient was likely to respond by rejecting not only the physician's advice but also the physician himself, as other studies on cooperativeness have shown (9, 10).

Observation also suggested that failure to follow advice may have been associated with inadequate information-giving by the physician. We indicated that one-fifth of the Panorio and shepherd patients understood nothing of what they had been told. Some unknown larger proportion probably were uncertain about one or another aspect of their condition or treatment. We observed that the physicians gave very limited explanations, did not check to see if the patient had understood them, and occasionally used medical terms that peasants were not likely to know. The peasants, of course, rarely did anything to correct the situation, being reluctant to "admit their ignorance" by asking the physician to explain further or to repeat himself. Some villagers displayed an attitude of quiet deference and social inferiority in the examination room but angrily denounced the physician as soon as they came out of the door.

Another factor that may have affected cooperation, but for which we have only our clinical impression as evidence, was the arousal of anxiety during the examination. We have indicated previously how anxious many peasants were about the role of the physician in giving the "idea" of illness, or in giving bad news. There is a considerable amount of anxiety in the life of the peasant, and it is expressed in his concerns over the unknown. This is clearly revealed in his preoccupation with the dangers of exotika, sorcerers, and persons with the evil eye, who may cause him misfortune. His anxiety often focuses on bodily functions, especially on malfunctions or pains that he cannot attribute to a mechanical cause, such as an accident, or to a well-known disease, such as an allergy or the measles. It is common for all humans to be anxious over matters that are both important and ambiguous, such as unknown physical ailments. Because of this anxiety, the peasants undertake a visit to the physician with much ambivalence, and, as our follow-up visits to Athenian specialists showed, when any superficially "practical" excuse presents itself, this ambivalence is resolved in favor of not going to the doctor.

During the medical examinations some patients became visibly more anxious: their palms sweated, they breathed more rapidly, they turned pale, and some came close to fainting. We had warned the examining physicians how important it was to reduce such anxiety by telling the patients what procedures they were using and what their findings were as the examination progressed; but our advice was not consistent with their own training or habitual approach, and they did not follow it. The doctor's grunts as he listened to the patient's heart worried the patient; when the examination was completed and the doctor said nothing about the condition of the heart (or of any other organ that had been found normal), the patient left with a new burden of anxiety.

Although we are hard put to explain the differences in cooperativeness among the three communities in going to Athenian specialists, there are two factors that may account, in part, for the higher rate of cooperation among the Saracatzani. The first factor is that one of the examining physicians was already known to a number of them, for he had a house in the country near the small village where some of the Saracatzani had built new, permanent residences. They had visited him there in his part-time practice; when they came for examination and found he was the doctor, they were quite pleased.* It may be that they were more willing to follow his advice because he was a familiar person who already had their confidence. A second factor in their cooperativeness may be the cohesiveness of the shepherds as a kinship community. The whole community was involved with the aims of the study because of the friendships we established with the leaders and their families. This involvement may well have led to pressures on them to follow up on the medical advice.

* Another illustration given is the case of the shepherd woman who went to a doctor in Athens for an eye injury. She was very pleased with the doctor, explaining that he was very kind and gave her special attention "since he came from shepherd people too." Shared backgrounds led to increased comfort and trust.

In reviewing the material in this chapter, we conclude that participation in medical examination is definitely not the same as following the physician's advice. Each may be a form of cooperation in medical care, but each is influenced by different factors. At each stage in the provision of care, one must attend to the variety of cultural, social, and psychological factors that reduce participation and cooperation, recognizing that the provision of adequate health care does not end either with the establishment of local facilities or with the undertaking of rational educational programs. Any newly introduced program must come to grips with the cultural milieu, anticipate its impact on various segments of the social structure, and be sensitive to the differing psychological characteristics of individuals within the population.

8

Hygienic Knowledge and Practice

We have implied earlier that health education is important in providing a foundation for better health care among peasants and shepherds, and we have indicated the magnitude of the educational task as we have presented information on health behavior and beliefs. In this chapter we present the results of inquiries about specific matters of hygienic importance: individual knowledge of body organs and their functions and hygienic knowledge and practices in the villages, as measured by Dodd's Hygiene Scale (18).

KNOWLEDGE OF BODY ORGANS AND FUNCTIONS

Questions in this part of the interview schedule resembled a school achievement test. They were not easy to present because respondents thought they ought to know the answers and were unhappy about their ignorance. We tried to be as gentle as possible, but when we asked the questions, it became evident to all concerned how very much the villagers needed and wanted better health education. Each family in a subsample consisting of half of the villagers in Panorio and Dhadhi and all of the Saracatzani was asked about a number of body organs and the blood. We asked what the function of each one was, what things were good for it, what endangered it, and what illnesses involved it. Since responses appeared to vary more within villages than between them, we present the results for all communities combined.

Villagers were questioned about the heart, stomach, brain, liver, intestines, blood, lungs, sex organs, and eyes. The majority of families were able to give a simple and roughly correct description of the function of the stomach, brain, lungs, sex organs, and eyes. But only for the eyes and sex organs did that majority exceed 75 per cent of all the families interviewed. Practically no one could state the function of the liver, the blood, or the heart.

When asked what endangered the organs, the villagers' most common replies for all organs were emotional distress, fatigue, certain bad foods, excess or overstimulation, and harsh environmental conditions (sun, cold, etc.). These attributions of illness cause are similar to those offered during the morbidity survey (see Chapter 4). The villagers mentioned that there were a number of specific dangers for specific organs. For example, they said that the brain is endangered by "bad thoughts about harming others," "too much thought," coitus interruptus, "unrequited love," and "overweening pride" (hybris); sex organs, they said, are susceptible to harm from a "bad life" and the use of contraceptive douches of lemon or quinine; and the lechona is harmed by drinking wine.

Although a number of endangering conditions were described, bacterial agents and structural defects were seldom mentioned, nor could the majority of families report any endangering agent or condition for the intestines, blood, liver, lungs, or sex organs. Because of the presence in Greece of a number of diseases involving these organs, some of them extremely serious, this meager knowledge presents a genuine hazard to the hygiene of the rural people.

We named the major body organs, plus the blood, for the villagers and asked them what diseases might afflict each. The majority of families were not aware of any specific diseases involving the heart, the liver, the sex organs, or the skin. Only a few ailments involving any organ were mentioned by more than a few families. These included heart attacks (infarct and angina), stomach ulcers, psychosis, headache, meningitis, enteritis, "twisting" of the intestine, jaundice, cancer and leukemia, the "blood's becoming water" (which is believed to occur in tuberculosis), tuberculosis, venereal disease, women's troubles involving the sex organs, pimples, eczema of the skin, cataract, infections and itching of the eyes.

Such a list indicates how few diseases, as defined by Western medicine, are known to the villagers in terms of organ sites of infection or disability. It also reflects certain misconceptions, e.g., that jaundice is a blood disease, and certain folk ideas, e.g., that the blood becomes water in a particular illness and that the intestine becomes twisted in another (the latter is related, at least in terms of folkhealing methods, to the wandering navel). Some categories in the list are rather broad, such as "women's troubles" or "infections" of the eye, which include styes, conjunctivitis, and trachoma; some are simply symptoms, such as itching.

The answers to the question of what conditions, agents, and therapies are recommended for the known disorders of the organs were consistent with what the villagers had already said. Recommended were emotional tranquillity, careful life styles (including self-care and the avoidance of excess and overstimulation from too much wine, smoking, exercise, work, or worry), pleasant or benign environmental conditions, protection from harsh environmental conditions, specific practika ranging from magic to alcohol rubs, special food and drink regimes (emphasizing healthful foods and avoiding excess, unhealthy, or "contrary" foods), hygienic precautions (cleaning and washing), harmonious interpersonal conduct (good thoughts, no fighting, chastity in women), medical care, and so forth.

RATINGS ON A HYGIENE SCALE

Some years ago Stuart Dodd, then at the American University of Beirut, developed a hygiene scale (18) that provided for interviews in and ratings of villages in the Near East. This scale gave a hygiene score for each village and family. The scores allowed one to interpret the status of a village with respect to other villages and Western hygiene standards.

We used the short form of this scale, having first revised it to fit Greek conditions. Unfortunately, we were not able to pretest our revision before going into the field, nor were we able to establish the reliability of questions and ratings that had been changed from their original form. But it was our impression that the revised scale worked well. It provided us with a means for assessing basic hygienic procedures and understanding in a subsample of families in each village, and to compare the resulting scores of one community with those of another.

Table 15 presents the total points scored by each village on each item rated. The nearer the actual point score approaches the maximum possible point score, the better the hygiene rating of the village. If the actual and the maximum possible scores were the same, the village would be rated perfect on hygienic practice on that item. The actual village score on each item is expressed as a proportion of the maximum possible score on it. Thus, on the first-listed item, diarrhea, Dhadhi is rated at 224 points over a possible maximum (perfect) score of 960. The maximum possible score changes as a function of both the item being rated and the number of families in the subsample, which accounts for the variations in maximum possible score by item and village. The villages are ranked for each item, with a rank of one being most hygienic and a rank of three being least hygienic.

After calculating the average per cent of actual scores over the maximum possible score for all items in each village combined, we find from Table 15 that Dhadhi shows an average of 50 per cent of the maximum possible score, Panorio an average of 52 per cent of the maximum possible, and the Saracatzani an average of 41 per cent of the maximum possible score. Because of the unreliability present in these ratings, one should not make too much of small differences. We do conclude, however, that Dhadhi and Panorio rate about the same in hygiene practices, whereas the Saracatzani lag behind.

TABLE 15

Scores Achieved on Selected Items of Dodd's Hygiene Scale by Families in All Three Communities

(Actual over possible scores)

	Dhadhi (N = 24)		Panorio (N = 13)		Saracatzani (N = 5)	
Hygiene Practice or Knowledge	Score	Rank	Score	Rank	Score	Rank
Diarrhea	224/960	3	133/520	2	64/200	1
Smallpox	72/282	2	168/260	1	49/100	3
Insect-proofing	45/360	1	15/195	2	0/75	3
Infant diet	2252/2400	1	1134/1300	2	237/500	3
Adult diet	445/480	1	234/260	2	79/100	3
Milk-boiling	430/480	1	180/260	2	40/100	3
Water purity	0/1200	3	550/650	2	300/300	1
Food-covering	105/120	1	40/65	2	5/25	3
Harm from flies	188/360	tie	104/195	tie	39/75	tie
Fly-breeding	144/360	3	120/195	1	21/45	2
Garbage disposal	110/600	3	65/325	2	85/125	1
Animals' access to house	300/360	1	135/195	2	30/75	3
Flooring	110/240	tie	61/130	tie	0/50	2
Sore eyes	32/480	tie	19/260	tie	0/100	3
Wound dressings	92/240	2	42/130	3	22/50	1
Average score	50%		52%		41%	

The relative similarity in over-all hygiene rankings among the three Greek communities should not lead us to overlook the dramatic differences among them in particular hygiene practices or knowledge. For example, Dhadhi is more often ranked highest on individual items of practice, and the Saracatzani are more frequently ranked lowest. The reason for the final near-equality of total scores between Panorio and Dhadhi is found in the abysmally low score that Dhadhi obtained for water purity: every well in Dhadhi lies within 30 meters of a latrine or privy.

Each community received a low score on the care and knowledge of diarrhea; even lower scores were obtained on practices and knowledge for sore eyes (including trachoma). Dhadhi and Panorio both made very poor showings on garbage disposal, for they throw their garbage into the fields or the ditch in front of their houses. The Saracatzani, on the other hand, burn their garbage, and received a much higher score on that account. The Saracatzani scored zero on three items: care for sore eyes, house flooring (all the huts are floored with hardened mud), and insect proofing. (The beehive huts of wood and straw have no means of keeping out insects; chickens and goats come into the huts, where the flies and mosquitoes are so thick that guests are covered with blankets for protection.) The high score the Saracatzani received for water purity is probably exaggerated. They bring water from an uncontaminated spring to the summer camp; however, in their wanderings as they herd the sheep, they drink from a variety of water sources, and one can be sure that many of these are exposed to contamination.

The high scores on infant and adult diets are untrustworthy. Here family members reported what they feed, or would feed, a baby under one year of age; for adults they reported what they had eaten during the previous 48 hours. Recall was often poor, and interviewers tended to ask questions that prompted affirmative replies. Replies were designed to please the interviewer and to make the family appear in a good light. Observations made of these same families at mealtime indicated that the protein diet was less than they had reported, as were the amounts of vegetables and fruits in the adult diet. The staples were bread, olive oil, wine, cheese, and vegetables in season that are homegrown or easily taken from a neighbor's field. Meat is usually eaten only on festive occasions.

The scores obtained for milk boiling were also falsely elevated. Observation and informal discussion led us to conclude that milk is often drunk fresh without being boiled. When respondents replied to the Hygiene Scale questions, they were replying, as Edwards would say (19), in the direction that is socially preferred; they reported practices they had learned they should follow, even if they did not consistently engage in them.

It is interesting to compare the hygiene ratings obtained from the brief form of Dodd's scale in our Greek communities with the ratings obtained in Syrian villages in the 1930's, when the Near East Foundation sought to improve rural health through clinical, educational, and sanitation programs. Dodd reports that the score for the control (peasant) village in Syria in 1931 was 241 points out of a possible 1,000 (24%); the score for a village chosen for experiment (education by a visiting nurse and mobile clinic visits) was 253 (25%); the score for a demonstration village (which had a clinic) of Armenian refugees with better education and more urban experience was 321 (32%); and the score for urban families in Beirut was 654 (65%). In 1933, after the education experiment was completed, both the control and the experimental villages increased their scores by 5 or 6 per cent. The demonstration village, in which the clinic team and its Armenian nurses lived side by side with the Armenian refugees and in which there was massive community involvement of the interested and intelligent villagers, increased its score by 207 points (21%); its total score was 528 points out of 1,000 (53%). Greatest improvements occurred in insect control, use of sensible medical remedies, and infant care. Very little improvement in food or housing standards took place, nor could it, in view of environmental and economic limitations.

Although certain of the ratings we obtained may be unreliable, the over-all scores achieved on the Hygiene Scale in our communities suggest that the level of hygiene in each of these villages is better than that of any of the Syrian villages studied in 1931. The level of hygiene in the Greek villages appears to be comparable to that achieved in the Syrian demonstration village after an intensive educational and health-care program was undertaken. But the hygiene levels of neither the Greek nor the Syrian communities approached the hygiene level of city dwellers who were, for the most part, educated, prosperous Syrian Greeks of the Greek Orthodox faith living in Beirut. Dodd's finding to the effect that clinics established within communities can produce dramatic changes in village hygiene suggests a means for improving the hygiene of the Greek communities in our study. Attempts to improve hygiene in the experimental Syrian village were not demonstrably effective under conditions where a visiting nurse gave advice but where no systematic program existed, no community involvement occurred, the educational level of the villagers was not high, and their adaptability was perhaps not as great as that of the Armenian refugees. According to the Syrian findings by Dodd and the Near East Foundation, public-health efforts among Near Eastern peasants are most likely to succeed when the peasants have a general educational foundation, are involved as a group in self-improvement, receive systematic educational programs, and have available in the immediate locale free medical facilities that are public-health oriented. Although it is not certain that the same requirements are true for rural Greece in the 1960's, one must keep them in mind.

9

Illness Interpretations: Natural and Supernatural

When illness befalls them, most men will ask, "Why?" It is a practical question, because if one knows the causes of an illness, one is a step nearer to its prevention and control. But it is also a speculative question, and in answering it men express their views of the universe: its nature, its forces, and man's role in it. During interviews in the morbidity survey, we asked each family to explain why each illness it reported for the previous year had befallen it. It soon became apparent that their replies could be identified and classified according to the concepts of illness they expressed. A few of these concepts—e.g., that an illness may have magical or supernatural causes—were only lightly touched upon in the first phases of our acquaintance with the villagers, but later on, during spontaneous discussions, much more information on these concepts emerged.

EMPIRICAL CAUSES OF ILLNESS

During the more formal morbidity-survey interview, we found that the most common answer to our question about illness causation was, "We don't know." We understood this to be, in part, an indication of the villagers' initial reticence to offer their ideas, especially ideas they feared we "city people" might ridicule. We also understood it to be an indication of their inability to fit many illnesses into any conceptual framework.

There were, of course, widespread individual differences. Some families were able to account for all the origins of the diseases they had experienced. Other families, often close neighbors of the people who could explain their illnesses, reported the same illnesses during the preceding year but could not account for the origins of any of them. These differences within the community are much greater than any differences between the communities. The percentage of "don't know" replies is nearly constant for the three communities: 31 per cent in Dhadhi, 31 per cent in Panorio, and 27 per cent among the Saracatzani.

Predominant among the explanations given for reported illness were the local living conditions and environment. The diseases thus accounted for were upper respiratory infections and afflictions of the joints, limbs, and extremities, such as arthritis, rheumatism, sciatica, and gout. People explained that they suffer from these ailments because they have too much work to do and get tired, because they are exposed to cold weather, because the heat of summer makes them sweat too much and then they cool off too suddenly, and because the humidity or climate is harmful. They also blame dryness, dust, wind, and sun for their disorders.

The emotions were often said to account for illness. This finding is similar to that of Kemp (37), who studied illness among the South Slavs. Among the people of Dhadhi, for example, worry is considered the exclusive cause of cardiovascular illness (heart pains and high blood pressure), and in Panorio and among the Saracatzani worry, emotional problems, and despair are said to cause stomachaches, headaches, dizziness (usually considered a symptom of high blood pressure by the villagers but sometimes named to indicate general distress or confusion), and, in one case, partial paralysis of the voice. Another frequently mentioned cause of illness was failure to take proper care of oneself. Here the villagers referred to carelessness, foolishness, and indulgence to excess. The indulgence most often mentioned in this respect was eating or drinking too much. Drinking too much wine was a primary reason offered for stomach troubles; other excesses were said to account for weakness, stomachaches, and pains in the limbs and joints. Taking too much of a remedy, such as mountain (herb) tea, was also said to cause trouble. The other extreme, too little food, was also mentioned. Illness was also said to be caused by failure to dress wisely, to work moderately, and to rest as needed.

The dangerous or unpleasant characteristics of substances inhaled or ingested constitute another source of illness. Bad food or water and unwashed food or "smells" were said to lead to poisoning, stomachache, intestinal disorders, and allergies. Heavy sauces (usually made with tomato paste and olive oil) and salty foods were considered particularly indigestible. Smells from the stables were said to cause such illnesses as allergies, although these smells were considered to have healing qualities as well.

Accidents and actions were also cited as sources of illness. Those mentioned included being knifed in a fight with a neighbor, being wounded in the war, being beaten by one's husband, hurting oneself by carrying something too heavy, and running a stick into one's eye. Such events usually result in visible injuries or wounds, but they may also be invoked to account for miscarriages, ear troubles (one villager blamed his deafness on his father's having boxed his ears), headaches, and dizziness. One angry woman laid the blame for her chronic headache on a beating she had received from her mother-in-law. One suspects the explanation was motivated, at least in part, by a vengeful desire to turn others against her mother-in-law.*

A change of environment is also regarded by the villager as a cause of illness. Visiting someone in another region or even traveling a short distance out of his own village is seen as sufficient reason for the sickness that befalls the traveler, either while he is on his journey or after he has returned home. It can be assumed that this explanation reflects the villager's fear and distrust of outsiders and strange places fully as much as it reflects his vulnerability to biological pathogens to which he has not developed resistance.†

One group of empirical illness-causing factors is noteworthy primarily because it was scarcely ever mentioned. For the approximately 350 specific illnesses reported as having occurred during the preceding year, explanations involving microbes, virus, contagion, and epidemic were offered only four times. After reviewing the reported illnesses, we estimate that a physician would assume the necessary presence of a bacterial or virus agent for about half of them.

A few villagers attributed their illness to the social and economic circumstances of their lives, saying they had been sick because they were poor or hungry or undernourished. Such explanations were more common in Dhadhi and Panorio than among the Saracatzani. We interpret these explanations not so much as reflecting an actual state of

* For a similar example of the social use of illness, see Margaret Clark's *Health in the Mexican American Culture* (15).

† See Phyllis Williams's discussion of "paesano mentality" among South Italians (73).

affairs at the time of our visit, but as expressing the uncertainty of the villager about his sustenance in a harsh and unpredictable world. Many families did not, in our opinion, have an adequately balanced diet; a very few may actually have been hungry. But when families who are not now hungry speak of hunger, we must assume that their preoccupation represents past hardship as well as anxiety over the future. This concern with hunger seems also to reflect the enormous importance of food and feeding in the rural culture and the preoccupation with the natural and symbolic effects of food and famine.

MAGICAL CAUSES OF ILLNESS

Magical diseases or disease causes were not often reported in the matter-of-fact discussions that took place during the formal interviews. During these interviews the "set" of the family was in the direction of the empirical. Only during casual conversations, when the villagers spoke more spontaneously, could we profitably guide them into discussions of magic. During these discussions it sometimes happened that a villager reported some condition that we had defined as an illness or disability but which he clearly did not regard as such; at least, he had not linked it with what he had taken to be our interests in the morbidity survey. Sometimes, moreover, we observed a disability that had not been reported and, asking about it, learned about the supernatural powers believed to have caused it.

The conditions that were most often associated with magical or supernatural origins include sudden swelling of the skin, some boils and pimples, paralysis of the voice or body, bruises, lesions and discolorations of the skin, and fainting or convulsions. These conditions were often attributed to the "exotika," a broad classification of the spirits of the supernatural world, including demons, devils, nymphs and neraides (female demons), revenants and the ghosts of the unburied dead, witches with and without form, and the spirits of the woods, caves, wilderness, trees, and streams.

In addition to the conditions caused by exotika, the villagers mentioned conditions caused by magical forces operated by other human beings. These fall into two categories. First, there are the discomforts caused by the evil eye, a form of witchcraft in which the bewitching person may not be aware of his power and may have no evil intent. Babies and children are especially vulnerable to the evil eye. Symptoms include headache, fretfulness, and malaise. A more complete list of evil-eye symptoms will be found later in this chapter. The second form of interpersonal magic that causes illness is not accidental; it results from sorcery, the manipulation of magical forces for the purpose of controlling or harming another. It can be used, the villagers believe, to bring bad luck and misfortune, injury, impotence and barrenness, wasting illness, and death.

INTERPRETATIONS OF HEALTH, ILLNESS, AND LIFE, AND RELATIONS TO GOD

Any serious sickness may arouse in the mind of the villager a consideration of his relation to God. This consideration will not take the form of a spiritual or mystic quest, for the ordinary villager is much too pragmatic and down-to-earth to be interested in piety that leads him toward asceticism. His response is that of a religious person concerned with what the illness reveals to him about his status vis-à-vis God. As our discussions with families progressed, we were able to explore some of the religious meanings given to illness. On two occasions it was said that illness comes as a result of God's testing or punishing the sufferer; but we inferred from what was said that illness is more commonly believed to result from one's failure to perform a traditional or ritual activity.

The villager is likely to contend that life proceeds more satisfactorily if one has the blessing of God, and that failure to earn that blessing increases one's vulnerability to misfortune. God rarely sends misfortune to the sufferer; rather, he withdraws the protective aura of his blessing and thus exposes the person to the ever-present menaces of the natural and supernatural world. But God's protection, even if it is available, is not complete. Like his ancestors, the Greek villager conceives of the universe as being inhabited by personified or deified powers of varying scope and capacity. The Christian God is the most potent of these forces, but not by any means omniscient, omnipotent, or omnipresent, as his church would have the peasant believe. God stands at the top of the hierarchy of supernaturals, but he is limited at best. Even with his help, one must do what one can for oneself. The same is true of the saints who protect man, as the following story illustrates: "A fisherman was drowning. He shouted out, 'St. Nicholas, help me!' St. Nicholas, the sailor's protector, answered, 'I shall, but you'd better start moving your arms and legs!' "

In considering how one must live if the power of God is to serve one, the villagers seemed to think that the pursuit of religious obligations, including those of faith and obedience to the Commandments, is the most important. Prayer and church attendance are not required, but keeping the sacraments and ritual purity are important. Personal decency is important too. Above and beyond the restrictions of the Commandments, the blessed man will be honest, will avoid stealing, lying, and gossiping, will give alms, and will perform his other duties for the maintenance of his human community. The belief is that God also blesses what the Greek admires in a man—honor, self-respect, and the enhanced philotimo of a worthy person.

How can a man so offend God that he loses his blessing and is exposed to the caprices of nature and humankind? A few say that God, like their parents, cannot fail to forgive, that no matter what a man does the divine blessing is forthcoming. But this is a minority view. People who have suffered, as all the villagers have, find it hard to believe that their lives are uniformly blessed. They are too aware of their misery and the cruel whimsy of the world. The man who invites disaster is one who, contravening the religious and social ethic, exploits or harms others, is inhospitable to strangers, kills or steals, disregards his religious duties, does not purify himself before religious rituals, is litigious, and lacks self-respect and honor, or, contrariwise, shows hybris, or has the arrogance to challenge God. Finally, a man forfeits God's blessing if he practices black magic, works with the devil, or—even less Christian—concerns himself not with the devil, but with using black magic to control the Powers in order to hurt or kill others.

The Greeks are an inquisitive people, seeking explanations for events as best they can. If illness or misfortune occurs, some find the explanation in the loss of God's blessing due to deviation from the prescribed codes. But as in all communities of intelligent men, there are those who disagree and are skeptical. A few of these dissenters claim that no one can know what God demands, since no one has talked to him and returned to report. A few others say there is no God at all, that only the saints and Mary exist, along with the spirits of the wilderness—the fraternity of the exotika. No one is irreligious, but some adhere to the old ways and recognize a pantheon in which the Christian God does not reign supreme.

In reviewing ideas of illness causation, we are struck by the contrast between the explanations offered by the rural folk in Greece and those reported in studies of patients in the United States (6, 10). Among the latter, serious illness is frequently personalized and attributed to the sufferer's own guilt. It is said to be the fault of the patient for some moral failure, in the Protestant sense. This attitude does not appear among the villagers; although they do attribute illness to their ritual failures, such failures do not imply a personal moral transgression. The closest they come to this moral self-blame is in the notion that God "throws illness" as a test, or that he may punish a community with an epidemic, just as Apollo punished the Achaeans for stealing the daughter of the priest of his shrine. God is rarely credited with the punishment of individuals. In a culture where maintaining philotimo requires that a man remain blameless, the peasant does not attribute his sufferings to his own sinfulness.

VULNERABILITY

In addition to those who are vulnerable because they have transgressed the traditions and Commandments that govern man's relation to God, there are certain persons who are vulnerable because of certain circumstances in which they find themselves. Most vulnerable are infants, pregnant women, women in childbed, menstruating women, others who are polluted by dirt, impurity, or recent sexual intercourse, and brides and grooms. The good, and those upon whom fortune has smiled, are also in jeopardy. There are also situations or conditions that increase the vulnerability of these individuals and, indeed, of any person. The changing, and especially the waning, of the moon is one of these conditions; night is another; and being seen by the moon or stars is a third. The sacred days of fasting, Wednesday and Friday, are dangerous, as is the noontime, when the "bad hour" is abroad. Of the seasons, spring is the most dangerous; September also presents special peril. Certain places are especially dangerous: streams and forests, and springs and wells, which are considered the realms of the neraides and the spirits. In addition, the "heavy" shadow of the fig tree is to be avoided.

SOURCES OF DANGER

Most frequently we found the villagers saying that dangers come from other human beings who wish ill will upon the infant, the mother, the bride, or anyone who is undergoing the rites of passage, the rituals and ceremonies that mark such important occasions as birth, puberty, and marriage. These evil intentions are expressed magically, most often in the form of the "binding" curse, which brings impotence, barrenness, or other disability upon those cursed. The binding curse, or "tying," as it may also be called, arises from jealousy, envy, long-standing enmity, or the desire for revenge. Sometimes sorcery is "for fun," in the sense that the sorcerer may enjoy doing ill for its own sake.

We have seen that the dangers which arise at the rites of passage come from persons who have reason to wish evil upon those who are acquiring new status and presumed good fortune. In this sense the vulnerable persons are not vulnerable because of an inherent quality of their new status, but because the evil in others is most stirred on such momentous occasions. At such times sorcery (a conscious effort to use magic to harm) and witchcraft (the use of magic without any conscious intent to do harm, as in the evil eye) are most intensely operative. Thus the vulnerability of certain classes of people to disability and illness means that they are prime targets for the ill will of others, and as such are more likely to experience disease, disability, and death.

If a person is incautious, he may also bring harm to himself through careless activities that direct the forces of magic against himself or a child. For example, the lechona brings fever upon herself by washing clothes in warm water, baking bread, or permitting another woman who has engaged in these activities to visit her. Here one finds the simplest bond of sympathetic magic between one form of heat and another, i.e., the baking and fever, a bond that is established or "tied" by the unwise acts of the vulnerable person.

Sometimes vulnerability was discussed in terms of emotional discomfort or social strain, without the imputation of magical effects. For example, it was said that when good things happen to a villager, the other villagers express their envy in gossip, criticism, and calumny. The villagers described their life together as an uneasy one, with each family feeling competitive and jealous toward any other that might achieve success or happiness. The phrases they used over and over were "We eat each other," or "They eat the bride," "They eat the newborn infant." Although at some deep level of the psyche such cannibalistic words may in fact reflect a desire both to destroy the persons envied and to incorporate what is admired (the dynamics that ritual cannibalism in primitive societies may reflect), the daily use of these expressions is more superficial and indicates only jealousy and hostility. To be the target of unkind talk is painful; the villager who is having a joyful experience is no doubt aware that this in itself makes him socially vulnerable.

One final interpretation of vulnerability was offered by the villagers. It is the idea that for the lechona and the newborn child the grave is open; for 40 days the earth is ready to receive them. This belief implies a knowing participation of nature in the cycle of human life; it also describes the health risk that quite realistically, at least in the past, has attended childbirth and infancy. Lack of medical knowledge about the pathogens causing childbed and infant mortality caused other explanations to be offered. We have seen that these explanations—witchcraft, sorcery, the spirit world, God, the heavenly bodies, jealousy, and competitiveness—are consistent with old traditions and the present themes in social and intellectual life.

BAD LUCK

Vulnerability to misfortune is related to the idea of "bad luck." As with so many concepts held by the peasants and shepherds, there was a remarkable diversity of opinion about bad luck. Villagers were far from agreeing on the nature of luck and the conditions that influence it. The most common contention was that bad luck is simply a notion used after the fact to describe an unpleasant event. A minority said that unfortunate events are decreed and inevitable. For them bad luck is a restatement of the fatalistic notion that events will transpire "as they are written."

Some noted, as they did in discussing vulnerability, that being a good person is a condition that invites misfortune. In making such statements, they are remarking on the human tragedy or the dilemma of Job—that being good does not lead to good luck and that in the larger scheme of things moral, justice is not for men. Others, when they spoke of bad luck, contended that it is God who sends it—or the devil. They implied that it accompanies the sacred; for example, they believe that the priest brings bad luck, and those who encounter him in the morning may very well return home because they do not wish to begin a day so inauspiciously. A few villagers believe luck is animate; they anthropomorphize it, saying it is a willful spirit called Vittora, which is attached to a man until he is about to die; then it flits away to find another man with whom to live.

The ways of maximizing good luck range from traditional magic to shrewd practicality. Those who pride themselves on rationalism say that luck is simply a matter of brains and good judgment. Others believe it is more than that, saying it can be carried and transmitted. For example, if on the first of the month a carrier of good luck enters your house, your welfare during the remainder of the month is assured. On the other hand, should the visitor be a carrier of bad luck, he will transmit to you a burden of misery and disease. The same idea is associated with the serpent; whether he is a house god or a passer-by, he may shape your welfare by the emanations for good or ill that his presence brings.

One notion of luck is associated with ritual observances and the avoidance of offense to the supernaturals. As long as one does the correct things and follows the traditions, one is relatively safe. But if one ignores rituals or taboos, including those that have over the centuries become meaningless to the participants, troubles are said to come. For example, if one gives vinegar to another after dark, urinates in a stream, gives away the first bread after harvest, allows a stranger to see a lechona, or swears, using the name of Christ or the Panaghia (Virgin Mary, the "All Holy") one can anticipate misfortunes that may take the form of accident or illness.

There are acts that can ward off imminent bad luck. For example, if one spits as one expresses admiration for a baby, one can ward off the distress the baby would otherwise suffer from the evil eye. Touching wood or making the sign of the cross and saying a hasty prayer can save one from injury at the hands of a demon; spitting on seeing a stranger or a priest protects one from the danger they carry. Spitting is a fine rural art and men take pride in their range and trajectory; but the spitting itself poses a public-health hazard among a population prone to tuberculosis.

Prophecy and luck are related within the context of decreed events. The dream of the mother may foretell the luck of the child, or again, at the time of birth the three Moirai (the Fates) may gather—just as they were said to do in antiquity—to decide what the infant's future will be. If it is the child's destiny, his "luck" foretold, to become ill or disabled, these misfortunes cannot be forestalled. But those who speak of the Moirai do not fully believe in the predetermination of events, for in spite of fatalistic expressions, the average Greek tries with every magical means at his command to influence the course of his life. Nevertheless, the existence of fatalistic beliefs inhibits protective action in matters of health and disease, for it is in regard to these that such beliefs are most common. When the villager sees little relation between what he does and what happens, as is the case with some health matters, he despairs of achieving any better state through his own actions and gives himself over to fatalistic acceptance. This failure to see a relation between action and health is partly due to the lack of health knowledge among the peasants—they do not know that a latrine next to a well may lead to typhoid fever. Their fatalism is also partly realistic, for even the knowledge of proper nutrition would not make it possible for a poverty-stricken peasant to purchase healthful foods for his family.

THE EVIL EYE

Throughout the Mediterranean world, and extending northward into Europe, one finds a belief in the evil eye. An ancient belief (20) that has persisted for at least 4,000 years, it attributes magical power to the human eye. This power may be thought of as magnetic, or as being the consequence of admiration or envy of the beholder for what he beholds. It may also be equated with electricity or emphasize the evil intent of the beholder, in which case its use approaches sorcery. Ordinarily, as we have indicated earlier, the harmful effects of the evil eye occur without any intent on the part of the beholder, and may often occur without his being aware of any special power within his eye.

Among peasants and shepherds it is commonly believed that anyone can have the evil eye. This person need not stand in any special relation to the bewitched; for example, a mother may bewitch her own baby, a neighbor may bewitch a horse, a stranger may bewitch an automobile, and a shepherd may bewitch his own sheep. The Saracatzani tell the story of a shepherd whose sheep died, one after the other. The shepherd sought to learn the reason. Suspecting that he might have given his sheep the evil eye, he put his guess to the test: He closed one eye and looked at the sheep. Nothing happened. He closed the other eye and looked again—one sheep died. Then he knew; it was his own eye that had killed. Horrified, he tore it out.

Some believe that persons whose eyebrows run close together or whose eyes are blue are more likely to have the evil eye, but these characteristics are not too important. As one villager said, "All of us

Greeks have hungry eyes, greedy eyes." There are those who can, they say, "knock a donkey off a trail at 500 paces," as if they were marksmen using guns. There was, in both Panorio and Dhadhi, a remarkable unwillingness to name those individuals who had a reputation for the evil eye. Their names were mentioned hesitantly, if at all, as though this in itself was dangerous and could harm the informant.

The effects of the evil eye are many, ranging from minor discomforts to death; it is capable of disabling targets as diverse as humans and locomotives. The signs of having been bewitched are headaches, dizziness, sleeplessness, weakness, fretfulness, numbness, yawning, vomiting or other stomach distress, hiccoughs, fever, sore throat, chills, inexplicable injuries, paralysis, inability to stand, physical deterioration, loss of appetite, tremors, or languor. The evil eye may cause the bursting of men or animals, the wandering navel, tears and sneezing, pressure on the heart, miscarriages, and even the death of men and animals.

In comparing these signs and effects of the evil eye with the signs and symptoms of illness, we find a remarkable correlation (see Chapter 4). Most of the signs that the villager uses to define the presence of an illness are the same as those he attributes to the evil eye. For minor ills, or for the beginning of major ones, it is likely that the peasants and shepherds will simultaneously define the presence of an illness and diagnose it as being the result of the evil eye. We had the impression, from observation and informal conversation, that the evil eye is the most likely family diagnosis for bodily signs that coincide with those used to define an initial stage of illness but which in themselves are not yet sufficient to allow a more specific disease definition. We suspect that some of the "We don't know" replies to our early questioning concealed the suspicion that certain diseases were the result of the evil eye.

CONTAGION

Most villagers know something about contagion, and although they rarely suggested it spontaneously as an explanation for illness, inquiry revealed that they are aware of communicable disease and take steps to protect themselves from it. During the formal interview, the villagers were asked which diseases were contagious. Most families knew that tuberculosis is contagious. A majority realized that upper respiratory ailments and the usual childhood diseases are also contagious. Smallpox was mentioned by a few, and one or two families spoke of leprosy, trachoma, typhoid, diphtheria, malaria, and dengue fever as contagious diseases that have occurred in the region. In speaking of the childhood diseases, some termed them "compulsory," adding that they wanted their children to have them and would expose them at the first opportunity. (The usual childhood diseases are considered minor affairs, and the sick children are not usually put to bed or restricted in their activities.)

An interesting concept associated with contagion is that any illness is contagious if the blood of the sick person is similar to the blood of one exposed. None of the villagers were able to expand or amplify this belief; they seemed simply to accept it. They denied that similarity of blood is a matter of kinship but insisted that it is a matter of the composition of the blood, which two persons from different families might have in common. While anthropologists would probably attribute the origin of blood-similarity ideas to family-kinship notions, folklorists such as Garnett and Stuart-Glenn (28), state that it reflects a feeling of solidarity among members of the same tribe as well as those of the same family. Garnett and Stuart-Glenn observe that the solidarity also implies mutual influence, and that the identity of blood is equated with the mutual or sympathetic influence of all parts of the natural whole. What happens to one person happens or is known to the other person because they are both parts of the same larger unit, which is characterized by solidarity and sympathetic integration. A folksong, after telling of a young man imprisoned by his enemies, says

> And as his father sat at meat, away in Babylonia,
> The wine as he was drinking it, turned turbid in the wine cup.
> "Now know I this day my son within a trap is taken."

In this song the wine and the blood are the same; the wine is the blood which, turning turbid in the son, turns turbid in the father's wine cup too.*

The idea of blood similarity reflects an orientation in which man is at one with nature; all aspects of man and nature are integrated in one "dynamic" system, an attitude that Redfield describes for primitive peoples everywhere and which can be presumed to be a heritage from

* The concept of the evil eye also implies, according to Elworthy (20), a mutuality and a sympathetic extension of power throughout the clan.

pre-Mycenaean times. In addition, blood similarity implies the illogic of magic whereby sympathetic effects occur in objects sharing the same external characteristics or names. Finally, the "similarity of blood," echoing as it does the social solidarity of family and tribe and community, indicates the shared identity of persons who are themselves not separate and who do not define themselves either in actions or in self-conceptualization as wholly apart from the other people who are significant in their lives.

As implied in the notion of blood similarity, there are a number of families, although by no means a majority, who believe that any disease is contagious in the sense that when one kinsman falls sick the others will also succumb to the illness. It is conceived that the disease which afflicts one person afflicts the larger group of which that person is a member—the clan or community that share the "same blood."

Another concept of contagion, one we have met before (Chapter 4), is that the "idea" of an illness is contagious. Any illness will be "caught" if the person has the idea of that illness, for the idea, in its impact, is the same as a microbe or the evil eye. The word is the thing; as the semanticists would put it, the term is reified, and with the expression of belief that the illness is present, the illness will materialize. It is believed that death can result from catching the idea of illness. The following account by a Dhadhi woman is illustrative:

> My mother's uncle had tuberculosis. When he died one of the neighbors, not knowing what to do with his clothes, put them in the storeroom. A day after the death the neighbor's mother went down there to look for something and saw the clothes. She was frightened at the thought that she might get the disease from that, and she got the idea [of the disease] in her mind. She went to the doctor, who told her there wasn't anything the matter with her, but it was too late, for she had the idea. And the doctor said that once a patient has the idea in his mind, it is very hard to take it out and cure him. And she died from that idea, just like it was a microbe, like jealousy.

The most frequently mentioned diseases considered by the villagers to be noncontagious are cancer and the cardiovascular disorders (high blood pressure, heart failure, etc.). One or two families spoke of malaria, stomach disorders, flu, pneumonia, hernia, head colds, tonsillitis, echinococciasis (infestation with cysts of the dog tapeworm), allergies, arthritis, diabetes, certain eye disorders, chicken pox, pleurisy, anemia, bronchitis, typhoid, the evil eye, fevers, and appendicitis. One Saracatzani family expressed a point of view, which can also be inferred from the remarks of other rural folk, that there are no contagious illnesses. They said they had seen people exposed to diseases that doctors said were catching but these people had not become ill. On the other hand, they had seen well people become sick without having been exposed to illness. As practical people who draw conclusions from their observations, they had decided that contagion is a false idea that doctors have! They seem to have been told, or at least to have understood, that contagion describes a one-to-one correspondence between exposure and subsequent illness. Finding this untrue, they have rejected the entire concept.*

In reviewing the information about contagious illnesses, we find that there is no one illness which all village families agreed is contagious. Tuberculosis is recognized as such by many, as are the upper respiratory infections. Nevertheless, a few families insisted that the latter are never contagious. No diseases were mentioned as contagious that are not, but half of the families did list as noncommunicable some diseases that are in fact either directly communicable or transmitted through a host or carrier. This finding, plus the fact that approximately one-fifth of all families said that most or all diseases cannot be communicated, indicates a widespread lack of information about contagious diseases.

PRECAUTIONS AGAINST SPREAD OF DISEASE

Families were asked what precautions they took against any disease they considered to be communicable. About three-quarters of the respondents said they isolated the sick person and avoided contact with him. Half said they did not use the same eating utensils as the sick family member. A few families said they tried to keep especially clean, sometimes using disinfectants for this purpose. Our observations led us to believe that the foregoing measures are honored more in theory than in practice.

Some families spoke of the importance of not thinking of or mention-

* There is a lesson here for public-health education. An oversimplified presentation of health information that leads the intelligent observer to conclusions that he can see are false can result in the complete rejection of hygienic notions. Many of the peasants and shepherds are astute, keenly observant, practical people. Although they may find it difficult to understand abstractions and may prefer concrete either/or notions, there will be times when it is wiser to educate them slowly to complex concepts rather than risk their repudiation of educational efforts through teaching unrefined and consequently incorrect ideas.

ing the name of the illness as a means to avoid catching "the idea." One of the most common examples of this is the stricture against saying "cancer," even if no one has the illness. Instead it may be referred to as "the exorcised." This means of reference acts to ward off the illness through statement of magical intent at the same time its existence is acknowledged. Tuberculosis is spoken of in the same way. Both cancer and tuberculosis may also be referred to as the "bad illness." Interestingly enough, cancer is never named among the contagious illnesses, and is specifically listed as noncontagious. Nevertheless the belief that the idea is catching even if the illness is not directly communicable is more widespread than our statistics would indicate. This is an important differentiation; it shows that villagers may express a scientifically accepted doctrine while at the same time, on another level, they continue to hold and act on a strong and widespread magical belief.

Another word taboo is more elaborate. In the house in which there is smallpox, one is never to say "egg," because the Greek word "avga" is close in sound to the word "avgatizo," which means to multiply or increase. The fear is that the smallpox pustules would respond to invocation of the word by multiplying.

One of the special difficulties in avoiding tuberculosis, according to a few of the villagers, is that it is a "jealous" disease which willfully and with conscious animation seeks to infect those who are well. Stated in this fashion, the belief is much like the ancient concept of the spirits, or "Keres," of illnesses that were anthropomorphized. The "jealousy" of tuberculosis is sometimes given a different meaning: the person who is sick with tuberculosis is himself jealous of the health of others; his jealousy becomes the wish that others have the illness too. The intent is what is dangerous, for it may magically succeed in making others ill.

One of the magical devices for counteracting contagion is the use of religious rituals. These include asking the priest to bless one's house on the first of the month (rarely used in any of the study communities but more common in the nearby town of Doxario); praying for the protection of the saints, the Panaghia, or God; and relying on ikons, incense, talismans and amulets, and magic that deflects ill away from the person. Apotropaic rituals, such as spitting three times before mentioning a frightful word, are also used (the same ritual is used to avoid casting a spell with the evil eye).

A very few families said they use no precautions against contagion.

Their reasons included the belief that disease strikes only those with similar blood; if the sick person's blood is not similar—as is presumably the case with a stranger or with the refugees from Asia Minor who live among the Attica folk in Dhadhi—no danger exists. Others are more fatalistic and say that illness is God's will, or the decree of the Moira; one may as well be resigned to the inevitable. One family stated that with the advent of penicillin no one need worry about contagion since contagious diseases could now be cured. This comment reflects the dangers of an inadequate knowledge of medical science; some learning has led to dangerous optimism and unreasonable expectations.

WHO IS PROTECTED?

A few respondents indicated that they believe physicians are less likely to catch illnesses because of their greater knowledge of precautionary steps and because their wealth permits them a life style that protects them from hazards. No magical immunity is ascribed to physicians. One family said they thought priests were immune to illness because of the protection of God. The rest of the villagers rejected the idea of priestly invulnerability.

The priests themselves did not share the skepticism of the majority. All three of the priests in the region stated that they were invulnerable to illness because of the power of their faith. God and the cross protect them. The priests enjoyed using as an example the partaking of the sacramental wine.* They pointed out that all worshipers use the same utensil, including persons whom the priests know have communicable diseases. Nevertheless, the priests insist, they themselves have never become ill from this exposure. They report that neither has any communicant fallen ill as a result. The immunity of priests and communicants from contagion is a point of great pride and support for their religious convictions.

HOW TO STAY HEALTHY

The problem of protection against contagion is one aspect of the larger issue of how one must act to maintain health and life. In considering this question, one confronts primary themes in any culture. Only limited information can be derived from direct questioning of the peas-

* In the Greek Orthodox mass, partaking of the wine is not limited to the priest. All participants take a sip from a spoon.

ants and shepherds, for their replies are necessarily limited to matters to which they have given some thought and which are explicit rather than implicit in their lives. Insofar as the rural people do verbalize explicit beliefs, there are wide individual differences of opinion; nevertheless, these differences revolve around common central themes. This pattern of individual variability and heterogeneity, within which are found unifying life themes and shared sentiments, is characteristic of nearly every area of inquiry in the Greek communities we studied.

Nearly all the families agreed that one can best stay healthy by adhering to proper life styles and regimes. By this they mean that one should dress well; eat well; avoid cold, overwork, and overstimulation; be clean; and get enough rest. These basic rules for hygiene and living are sometimes subsumed under the general exhortation to be moderate: "Nothing to excess." Moderation in work and appetites is not enough; there must also be moderation in one's internal state so that one has emotional balance, avoids undue worry, and pursues both internal tranquillity and external stability. Man, some of them say, was put in nature by God to be the harmonious element; to achieve this harmony, he must himself be in balance.

There are some villagers who do not share the age-old moderate views of the rural folk. Some of them expressed the theme of fatalism that we have met before. What is written is written, what is prophesied shall be; life depends, they say, on the "oil in one's lamp," on the allotted "loaf of bread." For them, whatever their adherence to village conduct norms, moderate or otherwise, health is not something that they perceive as being under their own control. It is fated; "sickness is thrown by God." For them the rule prescribes, or so they say, that one must accept what comes. In opposition to this passive orientation are a few activists. These people say that one must be hygienic and modern; one must take care to have clean food and water, to seek clean air, and to avoid bad climates; and one must go directly to a doctor if one suspects illness.

Another theme is expressed in the requirement that man, in order to maintain life and health, must see to his purity and cleanse himself of inevitable pollutions. Before communion one must abstain from food, sexual intercourse, and misbehavior. Women must avoid the holy table in the church sanctuary; menstruating women—the epitome of pollution—and the lechona must not enter the church at all. Also polluted are the murderer, the woman after miscarriage or abortion, and the person who is unwashed. Pollution is a matter of public as well as private concern; the life of the community, as well as the health of the individual, requires that the rituals be observed.

One aspect of health maintenance is the condition under which one can benefit from healing efforts. Some believed that there are people who, by reason of fault or flaw, are not eligible to return to health. Others suggested that the chances to be healed depend primarily on the nature or origin of the illness or one's own constitution. On the other hand, those who spoke of pollution said that a state of impurity is the hazard to being healed. Their implication was that the polluted man, weakened as he is, can never hope to recover. There seemed to be less agreement on conditions for the successful healing of the polluted woman: some considered her recovery prejudiced only because she is weak ("delapidated," as one woman accurately described her own post-partum condition). Others thought that the fear in which man and God hold her blood and her sex indicates that her danger lies in her own primal powers, which may go awry.

Whether one speaks of the sins of the parents or of their pollution's being visited upon the generations following, the belief exists that the actions or condition of the parents can account for the sufferings of the children, including the failure of the children to benefit from healing efforts. One family, adopting Christian concepts, said that the sins of the parents—for example, the drunkenness of a father—will be reflected in the child's poor health. Another family said that the child's poor health is the result of the pollution of the parents, citing the example of a child conceived during a period when the parents should have abstained from intercourse.

Other observations were made by the villagers about those who fail to respond to healing efforts. For example, it is believed that the mental state of the person determines his response to treatment. One who has the idea he is ill, whether it be a conviction or a hypochondriasis, will not respond. The person who is upset or jealous or grieving is said to "dissolve" and, by implication, to lack the will or intent to recover. Other families observed with some bitterness that good people suffer and die and bad people live well and stay healthy. Their awareness of the moral injustice of the universe, in opposition to the Christian credo that goodness will be rewarded, underscores the doubt and conflict felt by many of the rural folk about what life is and what it ought to be.

10

Views on Medical Care

In this chapter we shall consider the views of the local physicians, pharmacists, and trained midwives in the Doxario region with respect to the health practices and beliefs of the villagers. We shall also present the villagers' views about the use of physicians, discuss some of the doubts and problems the villagers express, and point to some of the factors that appear to influence what they expect of physicians.

THE PHYSICIANS SPEAK

Dr. Marietis, born and educated in another region, has been practicing in Doxario for ten years. He is now 45 years old and lives in the second finest house in the village, a house that contains his clean and simple office and a small waiting room, both located on the ground floor. He works hard and is respected by the local people. Although he is in constant contact with the people of Doxario and is certain to see at least a few of the villagers in Dhadhi and Panorio during the year, he was able to tell us remarkably little about the folk beliefs and practices of the region. As he said:

Only a small minority use the pre-Hippocratic methods. These few go to the "iatrissa," an old woman who knows how to boil herbs or treat wounds. Now most have a full understanding of health, for we doctors keep informing the people of the risks they run by using the old methods or neglecting health. There are almost no practika today, and no magicians, witches, or komboiannites. There used to be a komboiannitis here, but after doctors brought him to the police he quit practicing. . . . I doubt if more than 5 per cent of the people use any herbs or plants or charms nowadays. There aren't any pathological conditions that villagers don't recognize as such; oh, perhaps an individual case, but now when a doctor tells a person he has a disease the person believes it. Sooner or later the sick one will come in and a doctor will diagnose his illness. They do, of course, believe in the evil eye. I don't; they do because it is in the Gospel, where St. Paul speaks of it, and the majority think it exists.

One can attribute some of Dr. Marietis's remarks to an understandable desire to impress us with the modernity of the local people, as well as to his natural pride in the progress they have undoubtedly made with the help and teaching of the local physicians. Dr. Marietis is conscientious; he works hard to improve health conditions. We must not be surprised if his philotimo colors his report about the local situation.*

There seemed to be less distortion in his appraisal of the most prevalent local pathological conditions. The clinical cases that most often come to his attention are much like those reported in the morbidity survey and found in the clinical examinations. He reported frequent cases of rheumatic-arthritic-gout disorders, upper respiratory infections, injuries and accidents, and gastroenteritis. He attributed gastroenteritis, which is prevalent in children during the summer months, to malnutrition, saying that nursing babies are not fed while the mothers are out working in the fields, and that in general the older children do not receive proper food. These conditions come to his attention only when a patient suffers such perceptible symptoms as coughing, high fever, or pain. With ailments that are mild or that have a gradual onset, the people are less likely to consider themselves in need of medical care, and they will not come to a doctor until there are serious complications.

One of the problems Dr. Marietis faces is that the family and friends of sick people tend to become overanxious. The patient with a minor ill is also excessively worried, and consequently demands or expects too much from the physician. Patients do not seem to understand that recovery depends on factors other than simply the physician's intervention. Dr. Marietis commented on this:

It is because the patient so fervently believes in the doctor's power that he fails to see that the doctor is not the only force in the cure. One result is that the patient is deeply disturbed when a cure fails to come. It is a great burden

* Moreover, it is quite common for members of an educated elite to deny the folk traditions among their own people. Their denials are understandable insofar as the elite are newly arrived and wish to cast off any taint of superstition in their own lives, or are afraid that the observer will think less of them for having their origins in a traditional folk culture (57).

for the doctor. On the other hand, most patients don't expect miracles from a doctor. They respect science, of course; now that they read about science they expect more of medicine, but not miracles. Along with this, the tendency to use magic or to turn to the church has faded in recent years.

Dr. Marietis finds that if a patient has extreme faith in the doctor, difficulties are created for both doctor and patient. Nevertheless, he says he would not have it otherwise: "I believe that the faith in the doctor has more therapeutic value than any medical treatment." We suggest that at least one psychological activity that underlies a physician's reliance on the power of the patient's faith is magical thinking. Those who believe in magic assume that an event may be brought about by wishing it. If a physician believes that the patient's wishes are powerful, and if the physician thinks of himself as capable of directing that "power" within the patient so that a cure is effected, then the physician is being a magician. Furthermore, the physician himself "intends" for the patient to recover, and he may believe that these intentions, without any intervention in the form of treatments of known efficacy but perhaps accompanied by impressive technological rituals, will also help the cure. These two sets of magical intent may very well join in the physician's mind to produce a very considerable faith on his part that important things are happening in treatment and that "faith" on the part of the patient is a central ingredient.*

The extent to which the villagers make use of the doctor and have the confidence in him that Dr. Marietis finds necessary for therapeutic success depends on social as well as medical factors. For example, just as in antiquity (64), the confidence that the villagers have in the doctor is based on the doctor's reputation for making a correct prognosis, even if it is not a profound or difficult one. Similarly, as in antiquity (64), his reputation is enhanced if he can make a diagnosis without taking a history or conducting an examination. Dr. Marietis observed that these two cultural requirements work a hardship on the physician as well as on the patient; nevertheless, they continue to exist. He also stated that the use made of a doctor depends on his personality, on his approach

* The possibility that physician and patient may share in magical thinking should not lead one to overlook the reality component in their observations. The evidence is clear that suggestion and placebos do produce profound improvements in some ill people (24). But it is necessary to keep in mind that placebo effects tend to be difficult to predict, ordinarily are short-lived, and may occur in only a minority of patients so treated.

to the patient and the family, and—very greatly—on his conduct in the community.

The villagers, he said, attribute a high status to the doctor. To maintain his position, he must remain aloof from them, not joining them in the coffeehouse, not becoming friends with lower-status peasants or shepherds, and indeed, not having any but formal and distant relationships with anyone in the region. The doctor must be known for his moderation and propriety. He cannot, for example, become drunk in taverns or on feast days.

In maintaining his aloofness, the doctor creates a social distance that insulates him from entrapment in an affect-laden network of obligations and from the easy transfer of patients' neurotic feelings, which might become the basis for unreasonable expectations. It is a distance that helps confirm and demonstrate his status in the community and a status in which he takes pride. But, as one might expect, there are counterforces operating to reduce that aloofness. We see an illustration in what one of the Panorio people told us one day during a casual conversation:

> Dr. Marietis came here one day to treat someone. While he was here we could see he wasn't feeling well. He said he had a headache that wouldn't go away. Someone had bewitched him with the evil eye. Well, knowing that, he went off to see Mrs. Sotiria—you know, the one who is so good at curing the evil eye—who lives here. She said the xemetrima, and he was cured. Dr. Marietis was very pleased, and he told Mrs. Sotiria that she was a better healer than he was.

The story turns the tables on the good doctor, suggesting that local healing is better than the imported scientific variety and flattering Mrs. Sotiria, who gained considerable status by her feat. This is no affront to the doctor, for everyone accepts the idea that the evil eye is outside the scope of illnesses which physicians know how to treat. It also shows that the doctor is not so aloof after all, for he is subject to the same bewitchments that affect the villagers and can be cured by the traditional magic that they use. Whether the story is true is a good question. Dr. Marietis must deny it, in any event, especially in view of his attempt to persuade us that all is modern in the countryside.

We do not doubt that the story could be true; the other physician in town admits to being impressed by practika cures. This is not uncommon; for example, one of the members of the public-health team told us he had gone to a folkhealer for the treatment of a sprain. Some

Greek physicians, just like the laymen, are eclectic about their treatment, drawing on whatever resources seem practical and on their own beliefs about appropriate cures within the various illness systems. In any event, Dr. Marietis is careful to ensure his estimable isolation. Throughout the months of our study, we never saw him chatting with other villagers or engaging in the usual coffeehouse-taverna life. He sat alone when he had his coffee, and he walked alone or with his family when he took his evening constitutional. He was courteous and aloof.

We discussed patient cooperativeness with him. He contended that 80 per cent of his patients do exactly what he tells them to do. He attributed their cooperativeness to their faith in him and in his ability. However, he mentioned several treatment problems. One is that patients may use medicine prescribed for one illness on the occasion of another illness without coming to the doctor for a diagnosis for the new ailment. He attributed this practice to their belief in the general efficacy of drugs. For example, patients may use quinine for any one of many disorders under the assumption that if it is good for one illness, it is equally good for another. Dr. Marietis said that this kind of self-medication reflects the confidence peasants have in their ability to diagnose and treat themselves. A second related problem that he mentioned is the villagers' unwillingness to spend money on medical care if they can possibly avoid it.

When we asked about the economic problems affecting medical care, Dr. Marietis stated his belief that no more than 20 per cent of the villagers in either Panorio or Dhadhi lacked the funds to pay for medical care. His own fees, $1.35 for an office call, $3.35 for a house call to Panorio, and $5.00 for a house call to Dhadhi (he did not say what he charged for night calls), were not, he felt, so high as to work a hardship on anyone. For poor patients he reduces the fee, and for wealthier ones he raises it. He did not discuss his own economic position. Others in the village said he did not make much money, and that his comparative opulence resulted from the rich dowry brought to him by his wife.*

Dr. Pantos, the second physician of Doxario, was born in the town and has practiced there for many years. He is 65 years old, gentle, friendly, and talkative. He lives in an older house with a lovely garden,

* Dr. Marietis has one of the few automobiles in the Doxario region and is the only physician able to make house calls to the outlying hamlets. The other physician does not have a car.

which is his joy. His tiny office is upstairs, crowded with accumulated books, pictures, medicines, and mementos. Dr. Pantos had much more to say about his life and about the people of the region. He differed from Dr. Marietis in many ways, primarily in his willingness to criticize the conduct of the local folk. He insisted that the peasants and shepherds used the practika extensively, that the old women of the village were the first to see and treat the illnesses, that they did call in the priests as healers, and that their ignorance of hygiene was serious:

The people here have no hygienic conscience. They smoke, drink wine, and eat heavy food (salted fish, fatty meat, sauces, and fried dishes). From my experience I think that the people from Asia Minor are cleaner than the natives. Generally the villagers keep neither themselves nor their houses clean. Almost every house is without a water closet, and the situation in the public W.C. here in Doxario is simply beyond description. They dispose of their garbage in any way that is convenient to them, at the expense of private and public health. The flies constitute a permanent menace during the summer. The villager's [only] object is to eat well and to dress nicely—there are 15 dressmakers [part-time] in Doxario with whom accounts are settled more quickly than with the doctor or pharmacist, just so they can parade proudly in the square on Sunday.

The family as a tradition is on the decline. There is nothing to attract the members to the house. The father works and goes to drink in the tavern; the mother works in the field and leaves the housekeeping responsibilities on the shoulders of a small girl.

[Health problems are serious] partly because the state does not provide a [free] public-health doctor,* and partly because the representatives of the state who have public-health responsibilities [the regional public-health inspectors and doctors] are not effective.† The only reason for the low frequency of serious diseases is our [good] water and hard work.

Dr. Pantos listed as his most common clinical cases the upper respiratory ailments, gastrointestinal disturbances—which he estimated to affect 70 per cent of the population during summer—rheumatism, and other joint disorders. The patients come to him only after their illnesses have developed dangerously, having relied first on folk treatments.

* The Doxario region has no public-health physician at present. It is the policy of the Ministry to provide every region with the services of a free public-health physician, but implementation of this policy is, as can be imagined, difficult.

† What Dr. Pantos refers to here is apparently an accusation later made more specific by others in the community and the region: that regional and provincial government inspectors and physicians are bribed not to enforce sanitation and public-health laws that would entail additional expenses.

Sometimes they attribute infant intestinal disorders to teething and fail to understand that a disease process is present.

Dr. Pantos affirmed the idea that the confidence villagers have in a doctor depends upon his social status, his ability to diagnose without asking questions or examining the patient, and his ability to make an accurate prognosis. The physician, he said, must also maintain a reputation for propriety and high social status: "I'm Jesus Christ for some of them. . . . I'm dedicated to my family and to the profession, to a decent way of life. I am highly esteemed." But the esteem is very limited and unrewarding:

People don't appreciate the work of a doctor. They go to Athens and pay $10 to the professors for a visit, and the diagnosis will coincide with mine. And then they congratulate me on it, but don't pay me for my work. They cooperate in treatment, but the dark point is that they don't pay. The doctor is a "bad necessity." When time passes, I send a message asking them to settle their account; they not only fail to send the money but don't even send an answer. Then they won't even greet me on the street.

Regarding cooperation in treatment, he went on to say:

Not cooperating is a matter of their not having a hygienic conscience. . . . Many are ignorant and do not understand. It is my habit, when I give a patient instructions, to ask him to repeat what I just said. That way we both find out he did not understand me. This ignorance is a very difficult aspect of practice.

He discussed ignorance and its consequences:

They all use the practika, the xemetrima, the "remedy-wisdom"; charms and magic are in all villages. Or, for instance, the child may cry because he has an intestinal disorder or ear pain and they say, "Ah, it has fallen apart [become separated].* So they rub it so that the child will be put together again and the result is that they sprain its shoulder. One reason they bind [swaddle] the baby is for fear it will fall apart.

Then, of course, they will have a prescription for something and later give the drug to a neighbor who has, according to them, the same symptoms. According to the law nobody but the doctor can give injections, but here everybody *but* the doctor gives shots. I could report it to the police, of course, but I can't bring myself to; it is beneath my dignity. Even the pharmacists do it; for example, if a pharmacist doesn't have what I have prescribed, he gives some other medicine, or even some unprescribed injections.

* The word "komeno," which means cut off, separated, or fallen apart, also means ill-disposed or out-of-sorts, and peasant usage implies a physical substrate for an observed condition of distress.

In the old days they resorted to practikoi even more; now they more often employ the xemetrima. But nowadays, too, they are more willing to go to the doctor and even to the hospitals in Athens.

In discussing the economic problems of patients, Dr. Pantos said he was convinced that everyone in the region could afford medical care. He observed that those of his clients who were most remiss in paying their bills were often the wealthier landowners. "They even go from one doctor to another," he said, "in order to avoid paying for their most recent visit." Dr. Pantos charges only $1.00 for an office call, and from $1.80 to $2.70 for a daytime house call. After sunset the fee—which, he states, is regulated by law—rises to $5.30, and after midnight to $8.00. Because he has no car, he cannot visit the outlying villages of Panorio and Dhadhi.

His low rate of collections—no more than half of his patients pay their bills—is a source of great distress to Dr. Pantos. His income on paper is about $230 per month, but actual collections are only about $115. In his opinion all rural physicians would be much better off working for the government on a fixed salary. He thinks this is especially the case for younger physicians starting in practice, for the villagers have little confidence in the young and much prefer an older physician.

Many villagers turn to the established physician for help in nonmedical matters. Dr. Marietis said, for example, that patients come to him for advice in family matters. Dr. Pantos expanded on this, saying that they ask his help in reconciling family disputes, and in dealing with officials and the government, especially in making out forms and applications, and that they also ask him to serve as a (free) matchmaker. Dr. Pantos does not consider these nonmedical requests a burden; he is more upset by the demands that occur as part of medical practice. He is particularly unhappy about the fact that the patient—and all of the family members who accompany him to the doctor, as is the custom—immediately demand to be told the diagnosis, the etiology of the illness, and how long it will last. They are all overanxious, he said; consequently, they demand assurance that the patient will be cured. No matter what he says, there is a good chance that the family will troop off to another doctor to repeat the performance, seeing perhaps a series of physicians before deciding on the doctor who suits them.

The terminal patient poses a particularly painful problem for Dr.

Pantos. He said that most peasants approach death with dread and fear; when told of impending death, the family will weep and scream. One must be a hero, said Dr. Pantos, to be able to tell the family. Dr. Marietis reported only a few disturbances, but granted that some relatives seem to demand that loved ones should live forever. Consequently, it is only in the exceptional case in which the patient is an educated person that the physician will tell him he is going to die.

THE PHARMACISTS SPEAK

There are two pharmacists in Doxario, both of whom own small, one-room shops with an estimated stock of one or two hundred items. Both are pleasant men who expressed genuine interest in local health problems. Their evaluation of the most common local complaints corresponded generally with that of the physicians; they mentioned upper respiratory infections, joint ailments, and gastrointestinal disorders. One pharmacist noted that there is still some tuberculosis, ascribing it to poor nutrition and alcoholism. He added that in the village of Doxario there are several alcoholics with cirrhosis of the liver and several others with stomach ulcers.

Both pharmacists agreed that rural folk are not in the habit of seeing a physician in the early stages of an illness, and that they will first resort to folk medicine, knowledge of which is shared by all and exchanged freely. Villagers will come to the pharmacist to seek advice before they go to a doctor. The pharmacists try to avoid giving diagnoses, but willingly refill prescriptions and fill requests for medications for which prescriptions are not required.* In emergencies they also give first aid if they are convinced that the patient does not intend to go to a physician and that he would otherwise turn to practika. One pharmacist cited as an example of such practika the common local practice of putting manure on a wound; the other mentioned the use of tobacco or soot as a hemostat.

The pharmacists disagreed on the reasons the villagers are reluctant to go to doctors; one said that they are stingy, that they all complain they have no money, but that in fact, except for a few families in Panorio

* The most commonly purchased nonprescription remedies were said to be camphor and "pills" (Neotysinol, Butasalidin, Natralzin, and Norralzin), for rheumatism; aspirin, Algodin, and other similar preparations, for colds; cough syrups; and glycerine-iodine solutions, for rubbing.

and Dhadhi, everyone in those communities could afford medical care. The other pharmacist contended that most of the people in Panorio are too poor to get the medical care they need, as are from one-quarter to one-third of the refugees from Asia Minor who live in Dhadhi.

One of the pharmacists was aware that being ashamed of illnesses kept some people from going to a doctor. Venereal disease is one of the shameful ailments, and because of this people sometimes ask the pharmacist about it rather than a physician. Those with venereal disease are more likely to go to Dr. Marietis because his house, on the edge of town, is more isolated than Dr. Pantos's house.* In summary, it was the consensus of both pharmacists that the primary reasons for nonuse of the doctor are the prevalence of folk-healing methods and the belief people have—whether justifiable or not—that they cannot afford medical care.

In regard to folk-medical practices, aside from the exchange of folk-medical knowledge among the women, neither pharmacist was aware of any healing specialists in Panorio and Dhadhi. Both mentioned Kostas the Turk in Doxario, "whom the doctors were against," but who, said one pharmacist, was good at treating fractures. They knew of Vlachos, a folkhealer of national renown who practices in Athens, and of some famous healers and magicians in Euboea. One pharmacist was distressed one day when we met him, because an old woman from another town was in Doxario to sell healing herbs. He intended to meet this unethical competition by calling the police.

Both pharmacists seemed well aware of the folk-healing methods. They knew that the majority of mothers put charms on or about their children to avert danger emanating from the evil eye, the moon, etc. One described how literally the notion of "binding" (as in "binding curses") may be taken:

A pregnant woman came in [to the doctor's office] with her husband and her mother-in-law. She asked the doctor to examine her; he did, and found that she was ready to deliver. The mother-in-law and the husband disagreed, saying that they thought it was too soon and that there was danger of a miscarriage. So saying, the mother-in-law asked the doctor if he had a lock available. The doctor asked why. The mother-in-law explained that she had

* The observation that shame and embarrassment operate to dictate the villager's choice of healer, and in some cases to prevent his going to a medical doctor at all, suggests that the location of clinics in rural areas should be at the edge of town so that access would be less subject to community observation.

already put one lock around the girl—it was around her neck under her blouse—and since delivery was so imminent, another lock would be needed to keep the baby inside.

THE MIDWIVES SPEAK

There are two midwives in Doxario. One is Pitsa, a pretty young girl who holds a degree in maternity nursing and who is assigned by the government to provide prenatal, delivery, and infant care for people in the whole Doxario region. The other midwife, Constantina, is an older and somewhat suspicious woman whom the doctors consider an untrained practika. She insists that she has had two years of maternity-nursing training, and in this contention she is supported by the pharmacists. It appears that she has had some practical training, but nevertheless the physicians do not regard her as qualified. She works privately on a fee basis, as opposed to the free service rendered by Pitsa.

Of the two women, Constantina is less well informed. For example, she says the pestilential and epidemic illnesses that she considers to be common "have no special causes." She recommends a diet for most disorders except for tumors, in which case she applies a douche; in "serious cases" she recommends a visit to a physician. As for preventive measures, she does stress cleanliness, good diet, and the advisability of seeing a physician. She is aware of the widespread belief of villagers in folk medicine, magic, and sorcery; she cited a recent local case in which a sorcerer tried to kill four people by burying their pictures in the graveyard after piercing the portraits' eyes with needles and writing curses on the paper. Nevertheless, she does not discuss any untoward consequences of these practices on the health of her patients. "I deal with family problems," she said. "They feel more free to ask me [than the government midwife] how to correct the situations in which their daughters find themselves rather frequently. The girls work in the fields [of the landowners] with boys who exploit their ignorance in matters of sexual relations." She gave an example:

I performed a Caesarian cut on a girl who was five months pregnant. By my doing a Caesarian the girl was able to maintain her chastity, and thereby I prevented a tragedy in the family—because the girl had five brothers [who would have killed her]. The husband, when she marries, will take the scar as an appendicitis cut. Generally, because people gossip a lot here, I ask most of my clients not to come to my house, but to visit me in the clinic where I work a few times a week in Athens.

By what she says, Constantina gives the impression that she is on good terms with a number of the peasant women and that, because they regard her as someone they can trust and feel close to by virtue of shared beliefs and background, they come to her with their various obstetrical problems.

Pitsa is a lively girl, ambitious, intelligent, and charming, but lonely and dissatisfied with her life in Doxario because she is a stranger there, and her family and friends live in a far-off region. She has been in Doxario two years. During the first part of her stay few women came to see her, for she was so young and thin that she did not seem sufficiently maternal or strong to inspire their confidence. Now that the women of the region know her better, they come to her for advice and care. They prefer visiting her, she says, to seeing the doctor, for they are ashamed and embarrassed in front of a man. In addition, as far as child care is concerned, the local people find the doctors expensive.

She estimates that about 40 per cent of the local folk cannot afford the medical care they need. Others who can afford it will not pay the doctor because they expect him to work without payment. Peasants consider money paid to a doctor as "wasted," and would prefer to spend it on clothes or other visible goods.* The physicians respond to nonpayment by not making house calls to patients with delinquent bills, and, while continuing to give needed office treatment, they make their displeasure clear. The other side of the coin, Pitsa said, is that the doctors may "skin" the patients: a doctor may ask the mothers to come in more often than is necessary, or, because he lacks diagnostic acumen, he might have to see a case three or four times before he can accurately diagnose it, whereas a more skilled physician could establish a diagnosis in fewer visits. By charging the patient for visits caused by his own ignorance, the midwife said, the doctor puts a heavy burden on poor families with several children, each of whom will need to be seen during the year.

* The idea that money spent on doctors is "wasted" is widespread among peasants. Thomas and Znaniecki (66) in Poland observed that peasants expected the land to produce all their necessary food, fuel, clothing, and shelter; when farm products are sold, the peasants do not expect to spend the money on living expenses, but on other needs. These are well defined in advance: fancy clothes, dowry, or luxuries—there are no city inhibitions about expenditures. To the extent that money earmarked for property, taxes, and clothing has to be spent on an unexpected need, as in the case of a medical bill, it is considered a misfortune. Thomas and Znaniecki contend that such thinking underlies the "stinginess" of peasants.

Pitsa agreed with many others when she commented that the women rely heavily on folk medicine, the xemetrima, charms and herbs, and religion and magic when they detect illness in their babies or gynecological disorders in themselves. Although she discounts the utility of these methods, she has herself been cured by practika; she told us how a wise woman saved her finger, which a doctor wanted to amputate when she was a girl. While one may not wish to believe in folk healing, such incidents, she said, make one trust some of its wisdom. She described how these folk practices are typically applied:

After the birth the woman and the baby must stay inside for 40 days so that the wind, the air spirits [Aerika], will not take them away. They also cover the babies so that no one can see them. If a child is bewitched despite these precautions—for example, if he yawns, cries, has no appetite, or is restless—they will say the xemetrima, putting olive oil and hot coals in water, crossing the child, etc. Of course, the infant probably has the colic, but they will insist it is bewitched. If the xemetrima doesn't work then they take it to church and the priest reads [prayers] over the child and puts his stole [petrachili] over it. [During the same period] the mother or grandmother will boil different herbs to give to the baby. And if this doesn't work they may take him to the doctor, but by then ten or fifteen days have passed.

She went on, talking about the pollution of the postpartum woman:

As far as the lechona's period of confinement is concerned, it is the same as the discharge of lochia; there is a bloody discharge for ten days, mixed blood and lochia for 15 days, and then the white lochia for 15 more. This is the time they think they are polluted and impure; nothing I can tell them can get them to stop holding on to old traditions. . . . It is worse with the older women; younger ones get out more. . . . Still, they are not supposed to visit anyone, to go to church, or even to go out of the house after sunset; the ghosts, the exotika, might harm them. If I visit them—they are not supposed to see strangers—I should not mention anything about sickness or death in the community.

But the biggest medical problems, she said, do not arise from these traditional practices, but from lack of information and sanitation:

The most common infant and obstetrical-gynecological illnesses come from the neglect of cleanliness and from ignorance. Really, what I should do one day is gather all of the women into the church and teach them the fundamentals of hygiene. But I have no time, and besides my salary is unsatisfactory.* While the older women are more ignorant, one has the special problem with young girls of taboo [on discussions or demonstrations of sexual hygiene].

* She implied that she is not paid enough to undertake such extra effort.

Economics and nutrition play a role, too. She said that many mothers cannot nurse their babies because they don't have enough milk, usually because their own diets are not good. Inability to nurse an infant, coupled with the unavailability of a wet-nurse, is the equivalent of a death sentence for the baby, for there is neither refrigeration nor knowledge of infant dietary requirements.

Moreover, animal milk may be given raw.

But sometimes the child dies. . . . If the mother has rheumatism, the doctors here may tell her not to nurse her baby. Well, even if the baby does badly, the mothers think the poorer the infant is at first, the stronger he becomes later. Then there are some mothers who won't want to nurse to avoid having their breasts become deformed. They wean their children in one day; the rest usually wean them at eight months by putting pepper or quinine on their nipples.

When they shift to animal milk, they should, of course, boil it. I know that the people in Panorio and Dhadhi tell you [the team] that they always boil their milk. Well, some do and some don't. Those who have only wood fires don't really have the means to boil milk. The same thing with diet; they tell you they eat meat when they don't.

As for the diet of the pregnant women, I try to have them control their weight, but there is no scale in the whole region on which they can weigh themselves. Scales are a great luxury.

Women in the region approach childbirth with great fear, but this fear can be useful, Pitsa said:

The women who have had a bad experience with a previous pregnancy are especially afraid they will die in delivery. They should be afraid. Only if they are will they go to a doctor. Actually many do die in labor, but that happens in the hospital as well as at home. We've been lucky; no woman has died in childbirth here in the last year, although some newborn have died.

Greek women make too much fuss about labor pains. Of course they hurt, but Greek women are overly sensitive. Labor is difficult partly because they don't exercise or make any preparation for pregnancy; there is no "psychotherapy"; so they are in poor shape; they are afraid and they fuss.

It is quite a scene. I think it is an opportunity for many of them; they swear terribly at their husbands while they are in labor, cursing them and blaming them for their pain. If a doctor is there, he curses right back, swearing loudly at them while the women go on screaming.

If a mother does die in childbirth, family vengeance may be exacted from the medical personnel, as in the old days of blood feuds when "an eye for an eye" was exacted.

We had a gynecologist come to this village once to begin practice. At first no one went to him; then he failed in two cases and had to leave the village immediately. It is the same for me; it is not only because they get better and more complete care that I recommend to the women that they have their babies in the hospital in Athens, but because my work here exposes me to special hazards. If I fail in diagnosis or treatment, my own life is in danger.

The failure of the women to cooperate in their own care is another important problem, one related to money, information, and the kind of relationship existing between patient and healer. Pitsa said: "The most important thing in getting them to cooperate is to gain their confidence; but even so, no more than half of the women do what they are supposed to. This week, for example, I have told ten different women to buy maternity girdles. How many did? Only one."

Although the women in the community are her main concern, they are not the only source of difficulty.

Speaking of what they believe! The wife of one of the men here has just had a baby. It was her third girl. He wanted a boy and considers her responsible. He won't speak to her now, not a word. I have tried to explain to him, telling him that sex is a male-determined characteristic, but he just won't believe it.

In her role as midwife, Pitsa has come to know a great deal about the frequency of abortions and hymen-restoration operations. We have discussed these events in Chapter 5, and will not repeat them here. We should record, however, that the efforts at abortion using folk devices are dangerous to the health of the woman. Pitsa gave us a case in point: "I saw a young pregnant woman who died in the hospital. She wanted to get rid of her baby and, following her neighbor's advice, broke the uterus by putting a stick covered with oregano into it."

Pitsa plays an important role in the community, being a confidante of the married women, secretly seeing the troubled unmarried girls in Athens to give them advice, trying to train mothers in infant care and feeding, and providing an important source of encouragement and referral to local physicians and to Athenian clinics and hospitals. As the only medically trained person in the region whose services are free, she sees many women who would not otherwise receive attention. On the other hand, her own work is restricted by the lack of interest and cooperation shown her by the local physicians. Dr. Marietis, for example, discounts the importance of the midwife, and Dr. Pantos echoes local prejudices by saying she should not be young. Pitsa's low salary makes it hard for her to live without a supplemental income and deprives her of any feeling of being appreciated by the province that employs her. Besides, she has no automobile, and since local bus service is very limited, she cannot visit most of the outlying settlements. As a result, she provides practically no services to the people of Panorio and Dhadhi, nor to any of the Saracatzani mountain camps.

Her problem with the two local physicians, one that we inferred, although she did not discuss it, is symptomatic of the general lack of cooperation among local and government medical personnel. The two doctors in Doxario operate independently of and competitively with one another; there appears to be no love lost between them. Pitsa operates independently of the physicians, although she does make an effort to get her patients to obtain medical care. There is no coordination of effort to provide services or education.

Pitsa was the only medical person to call attention to the region's need for a medical center: a building equipped with laboratory, examination and transportation facilities, beds for short-term treatment, and a salaried government physician to provide care, either without charge or under the developing rural insurance program. Both local physicians would apparently be pleased to be appointed to either a salaried post or the position of insurance-program doctor. Their own collections are so precarious that government or insurance salaries would much improve their financial position.*

THE PATIENTS SPEAK

The people of Panorio, Dhadhi, and the Saracatzani encampments are, in the main, pleased with their local physicians. They speak of them with respect. Some complain of faulty diagnoses, high prices, or careless treatment, but such complaints are few. The factors that they see as limiting their use of medical services do not involve dissatisfaction with the quality of local medical services or with the personalities of the doctors. Peasants and shepherds speak of being hindered by matters of convenience and cost. Seven miles along a rocky road is a long way for a sick child to be carried on a donkey. One house call at night, costing $8.00, represents over 10 per cent of the total monthly

* The financial position of the rural Greek physician is like that of many British physicians prior to the institution of the National Health Service in England.

income for some of the village people. For those whose income is mostly food produced and consumed at home, it is a higher proportion of their hard currency. Even an office visit at $1.35 is not an insignificant cost to subsistence-level families.

A common criticism is that physicians exploit patients financially. Above and beyond the initial high cost of care, some are convinced that they are overcharged or that unnecessary procedures or visits are imposed. These complaints echo Pitsa's contention that doctors sometimes "skin" their patients by seeing them more times than necessary to establish a diagnosis. Before accepting these complaints at face value, however, we should recall the remarks of both Doxario physicians that patients demand an immediate diagnosis without examination, and that some of them march from physician to physician—from oracle to oracle as it were—in search of confirmation of an original diagnosis, for proof of their original skepticism and distrust, or for a new diagnosis that they happen to like. The results are that the patients think they are being cheated and the physicians feel they are being subjected to ridiculous demands.

Returning to the comments of the villagers about factors that make them reluctant to go to physicians, we find them complaining that doctors are rude, a complaint that evidently arises out of what peasants feel to be their inferior status position; that the doctors fail to provide enough information about their diagnosis or treatment; and that visits can be dangerous to their health, since the physician can "put the idea of the illness in your head." These complaints, all of which are familiar from earlier chapters, indicate that the barriers to the use of local and Athenian medical services are the same ones that inhibited participation in medical examinations and reduced cooperativeness in following medical advice.

If a villager can do something to prove himself superior to the physician, either by outright trickery or by telling tales in which the doctor comes off second best, he may very well do so. For example, one of the shepherds told the following anecdote: "Two of my cousins decided to tell the doctor they were sick just to see how much he knew. Both of them were in perfect health and yet that doctor told them they had some illness. That's what *he* thought!" This story illustrates not only the conflict between folk and medical views but also the villager's pride in the superiority of self-knowledge over the physician's knowledge. One can also see in it the distrust with which the villager approaches the stranger, putting his credulity to the test.

Another illustration of distrust, directed at the city doctors and demonstrating both the peasant's failure to understand why doctors do what they do and his reluctance to consign the dying to a hospital instead of to the care of their own family, is found in the following account:

Some doctors don't diagnose diseases right away. For instance, there is a boy here who has the "out of here" disease [cancer]. The doctors thought it was his heart, but all of his insides are rotten. The boy [a man, in reality] wants them to open him again and operate on him, but the people around here tell him that if the doctors open him he will die. The people don't want the doctors to do their lessons on him. The doctors, on the other hand, want to keep him in the hospital so they can open him up after his death and make experiments on him.

A problem that we had not met before, at least directly, was expressed by a few villagers who claim to have heard of cases in which the physician did not wish to help his patients, either because he felt he was not being paid enough or because he was angry or jealous. They cite a story, one apparently well known among the rural people but without any historical evidence to substantiate it, to the effect that King Alexander's death in the 1920's was attributable to his physicians. The villagers tell how court plotters wanted the King to die because he was so popular and democratic. They enlisted the King's physicians in their plot, which resulted in his being poisoned. They tell other stories of physicians who killed the wounded of the opposing side who were brought to them for care during the recent civil war. But such stories do not convey any widespread distrust of physicians. They are regarded as isolated incidents. In the King Alexander story, for example, in which the physician has acted as a political man and used his medical position to achieve personal political purposes, the Greek listener will understand the darker side of man and the strength of political passion. The tale is a reflection on man rather than a criticism of the medical profession.

The villagers' awareness of the humanity of their physicians is also reflected in their acceptance of doctors' mistakes. Naturally, a mistake is unwelcome and can lead to antipathy between a family and a physician; nevertheless, most rural people expect no miracles of healing or perfection from the medical profession.* The following three comments by villagers are illustrative:

The *good* doctors read and study. My little granddaughter was sick and we took her to two different doctors, each of whom we paid well. Neither did anything for her. Then we took her to a third doctor, who, of course, neither could nor should have said anything bad about his colleagues. What he did say was that doctors do make mistakes and that some don't keep on studying. He gave us some new medicine and the child was cured.

A doctor can harm you if he wants and nobody will know about it. My sister was sick with malaria, and she began to urinate blood. A doctor gave her some injections, but they were the wrong type and she turned black. We went to him in the middle of the night; when he saw us he realized he had made a mistake, even before he saw my sister herself. He pleaded with us to feel sorry for his children and not to mention his mistake to anyone. He did his best and did cure her, but it could have been fatal.

Doctors aren't gods. They can't always cure you. God's blessing and His will are necessary. I remember when my brother died. We kept asking the doctor what could be done and whether there was hope. He told us our brother was out of his hands and was in the hands of God; only a miracle could save him. Another time my brother-in-law was badly burned, and some of the doctors told us he might not live. One of the doctors replied he would cut off his own hands if he [the patient] wasn't cured because his [the doctor's] science told him that the cure was possible. So you see, doctors aren't gods, but they do understand.

When, as the doctors have complained, peasants seem to demand assurance of a cure, even in terminal cases, it is their intense anxiety for the patient that motivates them rather than any conviction that a cure can be expected of the doctor. In this important regard the doctor-

* Compare the emphasis Dr. Marietis placed on faith as a prerequisite to medical healing and his view of his patients as people who have great faith in him with the matter-of-fact feelings that some patients describe in themselves. The basic inconsistency here may be the result of sampling different populations, for Dr. Marietis sees more patients from Doxario than from our three villages, and his patients may actually be more faithful. This inconsistency may be the result of interview bias. The patients may flatter themselves with a skeptical "I am a no-nonsense fellow" description and Dr. Marietis may do the same by saying how much faith his patients have in him. But it may also represent a genuine difference. Perhaps physicians need to see their patients as having faith and being cured by it; perhaps Greek peasants are more skeptical than the physician would want to know.

patient relationship in rural Greece seems to be on firmer ground than that relationship in more technologically advanced countries. Ordinarily, the peasant does not expect more than routine medical care from a physician. In the United States, however, many patients—more specifically, that unknown but probably large number of unreasonable patients—expect from their physicians love, promises of cure, and protection from the danger or the truth about their sickness; they want to be encouraged to false hope, and they demand miracles of healing. The American physician is expected to be father, priest, and magician rolled into one (10). But the peasant is deeply immersed in an extended, intense family; he does not need an extra father. He is immersed in an extensive and encompassing religious system and has more than enough priests around him. He is immersed in a magical system and can perform his own magic; if that doesn't work, he has priests, wise women, and magicians who can work magic for him. The physician need not perform these services or fill these roles.

Medical practice in Greece appears to us to be less contaminated with these particular needs and magic-religious thinking than in urban America. We cannot state this as an established fact, for our observations are limited and should be tested by further study. But our impression is that for the peasant and shepherd medical care is a rational, commercial human endeavor which one negotiates with city-trained specialists. As such, it is a limited and restricted business from which one cannot expect too much. Because it draws solely on human knowledge, it lacks the immense power of the natural-supernatural world. Medical science—for all the intellectual prestige it holds for the peasant—is a drab, humdrum tradesman's work that lies completely outside the dramatic, exciting, intimate sphere of religion and magic. Science cannot hope to compete with the ultimate powers, for it is apart from nature's way and nature's rhythm. Medicine is a technological product, no better and no worse than those who practice it; and those who practice it—whatever their status—are only men, and the peasant knows that as men they stand a poor second to the powers of life and death.

By virtue of the physician's great social distance from the patient, he is insulated from the easy transfer of fixed, inappropriate expectations. He is included, and rightly so, in the culture's generalized attitudes toward all authority (the wealthy, the educated, and the influential are powerful); he is subject to the standard manipulative ploys of gifts and promises that the weak man uses to propitiate the strong. (A later section will discuss this in greater detail.) But he seems to be relatively free of those particularly unrealistic expectations, seen in urban people, which confuse him with intimates, and which ask of him far more than the medical-treatment situation is designed to provide.

REALISM OF THE PHYSICIANS

The Greek physician sees himself in the same realistic light. He recognizes his limitations, and in a culture where there are already a host of supernaturals, including a God in which the physician almost inevitably believes, he is not likely to confuse himself with the deities. He may, as we have observed in Athenian physicians, strive for high status and power within the human system, but he does not bestow on himself the powers of the supernatural system. Unlike some American physicians we have observed (11), the Greek physician—certainly the Greek country doctor—does not assume the God-like role of the miracle-maker. If he wants magic, he can consult a magician.

A country doctor's office always has an ikon or two on the desk. Some have little shrines in the waiting rooms. Physician and patients are collectively part of the larger religious milieu, and both seem to keep in mind their subordination to larger powers. The physician may, of course, draw on that power in his work, saying of himself that he heals with the help of the saints or God. His patients will agree. Even though he may mediate the power, he does not assume the role of the priest in doing it, since there is already a priest in the village whose powers are not to be challenged.

Unlike the physician in the United States, who is a member of the profession most respected, the Greek physician ranks lower in the community status system. The rich landowner and the teacher rank higher, as may the president of the community, the owner of the largest cow herd, or even the police chief. In Doxario the physician, though respected, is too narrow a specialist to rank highest. Moreover, because of his aloofness, he is somewhat peripheral to its status hierarchy. In the city there are many who rank higher: royalty, the powerful government officials, rich businessmen, church leaders, and the professorial elite. The physician's secondary rank is reflected in the patient's attribution to him of reasonable human skill rather than unlimited power.

CHOICE OF PHYSICIAN

In seeking medical care, the rural people are guided by a few primary considerations. Convenience and cost are two; the physician's reputation with family and nearby villagers is another. Being kin is a fourth consideration. This kinship is achieved outside of any actual blood or marriage ties (neither of the local doctors has any such tie with the people in the communities) through ritual adoption, expressed in the "koumbaros" or godfather system. When a baby is to be baptized the family selects a godfather, usually someone of greater wealth or status whose protection for the child and affiliation with the family are expected to be beneficial. Among the families of Panorio especially, the physicians of Doxario have been adopted as godfathers. Adoption produces mutual obligations: the loyalty and support of the family is rendered to the godfather, and he in turn is obliged to offer protection and assistance. Of the resulting relations one Panorio man said:

> We get along very well with the local doctors. Neither of them demands money when they come (unlike Vlachos, for example), but will let the families pay as they can. To the poorest they give the free medicines [samples] they may have. I am ashamed when some here do not pay the doctors. We get on well because we are related—in a way—to the doctors, for they are godfathers [koumbaros] to children or people of the village.

When discussing the necessary qualifications for the successful practice of medicine, peasants and shepherds stress the importance of intellectual capacity and interest in the field. Less frequently mentioned are such factors as education, experience, and the physician's own confidence in his work. Some families also indicate the importance of religious factors, saying that the doctor's success is a gift of God, implying thereby that the physician should be open to and deserving of being an instrument of God. One villager added the interesting observation that there are two kinds of physicians, "the ones who have a female brain, a brain that gives birth to ideas so that it can perceive more, and the ones who have a male brain. No matter how much *they* study, they cannot become very good." This observation suggests the conflict between the folk concept of the woman as life-giving, nurturing healer and the fact the physicians are male. The conflict is resolved by projecting a female quality into the successful male physician.

Although the religious and moral codes provide general requirements, being a healer demands special conduct vis-à-vis the patient. This is one of the few areas of belief or conduct in which medical doctors are not differentiated from folkhealers in regard to expectations and beliefs: the healing role is a general one, and all healers, regardless of the technique or illness system they use, must show certain basic kinds of behavior. Most important of these is the capacity to deal with the task at hand—to take the necessary and curative steps. Second in importance is the ability to give support and encouragement to help the patient face pain or danger. Third is the demand that the patient or family be told what is wrong and what must be done. Fourth is the requirement that the healer show sincere interest in the patient and listen to what he says. Courtesy and respect are also required, and although manners need not be polished or sophisticated (and, indeed, country people would find such manners disconcerting), frank regard for individual dignity is expected.

PATIENT OBLIGATIONS

The healer is not the only one whose conduct is prescribed by individual expectations and cultural demands. Patients also have obligations toward the medical practitioner. Three are mentioned with regularity by families: the obligation of the patient and family (1) to pay the doctor's fee, (2) to obey his instructions for treatment, and (3) to provide the doctor with the full information he needs and, conversely, not to conceal diseases, symptoms, or relevant past events from him during the history-taking. One Dhadhi man put the patient obligations this way: "You don't deserve the doctor's interest if you don't follow his advice or are whimsical about it. Good manners are important too; they attract a doctor to the patient, and when you don't tire him and don't insult him, he will have patience and respect for you. If you're in the hospital you shouldn't tire out the nurses either, and it's not right to be a tattletale and tell the doctors what the nurses do wrong. After all, it's the same as with teachers. They know what their duty is and what is right to do."

Sometimes respondents say that the obligations of the patient to the doctor are the same as those of the child to the parent. The child must respect, honor, and revere his parents, show good behavior by being honest and decent, and obey by paying full heed to parental regards, wishes, and commands. Although the patient's obligations to the doctor are modified to meet the restricted and purposeful medical situation, obedience, respect, and fulfillment of his side of the bargain are expected of him, just as they are of the child. Such conduct in the patient seems very reasonable from a rational medical standpoint, and it is also consonant with cultural themes.

Perhaps the parallel in the conduct required of the child and the patient toward respected authority figures sheds some light on one of the major complaints of patients about doctors—that they expect payment for their services. As we have noted earlier, the villagers do not seem to have unrealistic expectations of the physician's ability to cure illnesses or to solve all problems, but they are often unreasonable in their expectation of free care. The majority of peasants contend that the doctor is obliged to receive and treat them even if they do not pay him or otherwise fulfill their obligations to him. Their assumption seems to be that just as the parent is expected to care for his child regardless of whether the child fulfills his obligations to his parent, the doctor—another authority figure—is expected to care for the patient regardless of whether the patient fulfills his obligations.

Another factor that seems to contribute to the peasant's balkiness in fulfilling his obligations to the doctor is his long experience with healers who are part of the informal social system—witches, wise women, and hand practikoi—from whom he nearly always receives care without having to pay for it, at least in currency. Appreciation is expressed, gifts are given, and obligations are incurred, but money rarely changes hands.

With a background of such expectations of parents, other elders, and healers, it is not surprising that the patient hopes the physician will respond in a similar manner, giving care no matter how the patient fails in meeting his financial obligations. Of course, if the patient is seriously ill or disabled and thereby dependent on the doctor, one must expect that the regressive emotional component of the patient-physician relations will show more childlike expectations.

It is in this area of doctor-patient obligations that disputes arise, for the doctor sees his job in a technical or business-contract light as a transaction between strangers bound by joint obligations of an immediately reciprocal nature, whereas the patient hopes for informality and indulgence from the doctor, allowing for reciprocity at a much later time. Disappointment can be expected in both parties. Although the local physicians may not say it in so many words, they understand that they are caught up in the informal system; witness their willingness to give free care and to hope for later payment. City doctors, on the other hand, may be further removed from these informal bonds; therefore it is between the peasant and the middle-class physician of urban origin that we can anticipate the most serious conflicts arising from unrecognized differences in expectations in the doctor-patient relationship.

There is another factor here, one that was mentioned in an earlier chapter: a sense of noblesse oblige in the peasants' view of the obligations of those with wealth or relatively high status. In return for respect from the peasant, the elite are expected to offer service, just as the high-status godfather is expected to offer bounty in return for the loyalty of his adopted family. The rich landowner should provide the plot of land on which the school is built; the man of moderate means will be asked to give food to the poor man at a time of hunger; and the physician is expected to give what he can to those more needy and uneducated than he.

In this sense the doctor is expected to give a gift of service, and in so doing he is asked to do no more than others in the community. Thomas and Znaniecki (66), in studying Polish peasants, observed the importance of gifts as a means of expressing solidarity within the community, establishing between giver and receiver a bond similar in some ways to family bonds, and a relationship that is conceived to be of greater value than the gift itself.

But gifts have another function in rural Greece; the gift once given does indeed establish a bond, which can then be manipulated. One gives a small thing and receives a greater. The Greek peasant, not without craft, may put it this way: "Obligate the doctor according to what you can give him, with money or something else, and then, if he treats you well, 'with all its sorrows and troubles, life is sweet.' " That is, you have done your best and won at least this round. Or, "Take a gift to the doctors; obligate them with material things." To entrap through gifts, to ensure the greater service through rendering the lesser, is one way to handle physicians or anyone else with power.

But the doctors sometimes turn these gifts to their own advantage. We have heard them demand of patients that a gift be brought: "Why have you brought me no figs such as I saw you carry this morning?" we heard one say; or "Here I am serving you and you have not even brought me a chicken!" or "What a lovely place you have here! One of my relatives is coming to visit; don't you have an extra room for him?"

If he has no gift to give, the patient may beg. There is no humiliation in this, for the man who provides for his family by demanding goods or services conceives of himself as a proper—and perhaps even a clever—parent. The beggar is well aware that those whose aid he asks will fear his evil eye—the spawn of his envy and admiration—should they fail to give. It is the same with a woman who accosts the doctor in the street to ask free advice regarding the illness of her children. Her entreaty may not be without an element of threat. Who knows what she might not wish upon one who would deny her?

Annoyed as the physician may be, he is likely to acquiesce in such demands. Life is not so certain in rural Greece that one can afford to offend anyone. He who gives today may well be the beggar tomorrow. And many are the tales of the fall of the great to prove it. Demands, gifts, and begging are all in the context of mutual help, shared dependency, unequal need and unequal power, and not a little underlying anxiety over the future.

In accommodating requests, the physician will most certainly complain, but he too shares the conviction that status puts an obligation upon him, that his philotimo is enhanced by generosity, that the beggar may have magical powers to harm and to avenge bad treatment, and, in particular, that hybris will be punished. In the physician's case it would be overweening pride indeed to pretend immunity from the decrees of fate and, therefore, fail to give the poor some of the service they demand. A Greek proverb warns: "I have seen many men who are 40 years old, but I have never seen anyone who has been a lord for 40 years!"

11

The Extent of Folk Healing

In this chapter we shall begin to consider the folkhealers, those peasants, shepherds, and priests who are recognized as having information or skills useful in the relief of pain and the treatment of illness and disability. We shall see whom the rural people designate as their healers, learn what skills they are said to have, and examine what factors appear to influence the use of these healers.

Folk-healing practices are widespread and varied. They range from the exchange of advice among village women about which herb teas are helpful in treating a common cold, to the formal visit of an entire family to Athens or a nearby town for treatment by a well-known magician or a komboiannitis. Among the villagers the names of a few outstanding magicians and komboiannites are held in the highest regard; these men are much more often referred to by name—with awe—than are any physicians.

Folk-healing practices and practitioners have a significant role in the healing activity in the villages. The most common folk-healing activity is the day-to-day exchange that takes place among the women, and sometimes among the men, and between a family and one of the wise women of the locality. Less common is a more formal visit to one of the part-time healing specialists in the village, one who is especially powerful in her use of the magic for the evil eye or who can cure the wandering navel. Still less frequently used are the priests, whose healing role is important but complicated by the antagonism between them and the villagers. The folkhealers least frequently called upon are the highly reputed specialists who live in distant towns—men or women who are visited only in cases of dire need. The cheap and easy local remedies are tried first, before an illness has become a serious threat to comfort or performance. Once a condition has progressed, other treatment means will be applied, usually several at once, so that treatments involving several different concepts of illness causation are at work at the same time.

There are several factors that appear to influence the choice of healers. In the early stages of an ailment or minor disorder, ease, free advice, and the comfort of talking to a fellow villager dictate the choice of someone in the village. Ordinarily, the very first exchange of information is within the family itself; if the family has its own healer, she (or he) will take charge. For most minor disorders it can be presumed that the family healing lore will suffice. If there is some delay in recovery, or if the family has no woman with healing skills, and if we presume now that there is some anxiety—which among Greek peasants seems to lead to intensified sociability—the patient or someone in his family will seek the advice of a neighboring wise woman or other person possessed of special skills or knowledge.

If these efforts fail and if the home remedies and rituals that are being tried during this same period also fail, the family will begin to consider taking the more difficult, more expensive, and in some ways more anxiety-arousing step of visiting someone outside the village. The choice between folkhealer and medical specialist will depend not only on proximity and anticipated cost, but on the reputation of the alternative healers among the family members. It appears that past acquaintance and ties of kinship—for instance, the relationship to a koumbaros—are of some importance here. Of greater importance will be the illness condition itself. How it is diagnosed, what its causes are thought to be, and what beliefs are held about how such conditions are best cured will determine the particular choice of healer.

Those conditions attributable to the evil eye—and they are many—will already have led to the diagnostic and curative spells involving the magic words, the xemetrima, and the rituals of holy water, olive oil, and perhaps incense. Since the spells against the evil eye are all available to the wise women within the village, it is unlikely that a continuing illness so treated will still be attributed to the evil eye. Consequently, the diagnosis of the evil eye will probably not lead to a visit to an out-of-town folkhealer unless the conviction of magical cause is so strong that the family is willing to upgrade the presumed source to include the possibility of sorcery. If sorcery is presumed—and this is a most unusual and upsetting conclusion for a family to reach—a visit to a powerful magician will be in order. There are no such persons in Dhadhi, Panorio, or the Saracatzani settlements. The family will have to go out of town.

A powerful magician will also be sought out if the condition is ascribed to any of the supernaturals outside of the Christian system of the saints, Mary, Christ, or God. These supernaturals are many; they are grouped into the general classification of the "exotika"—the nymphs and neraides, the devils and demons, the kallikantzari, or hairy people, the "moro," the vrikolakes or revenants, the ghosts, the "bad hour," the "Vittora," and others. Although they are not exotika, the malign influence of the moon or stars or the heavy shadow of the fig tree may also require strong counteracting magic.

Fractures and sprains and some cases of arthritis and rheumatism are the business of the komboiannites. These are always men because their work requires strength, not subtlety or tenderness. They also treat certain diseases that "doctors do not know," such as the wandering navel, the waist, and jaundice. If the shepherd or peasant diagnoses these ailments and, as in the case of fractures, finds that local healers are not skilled enough, he will choose an out-of-town healer.

There are also certain occasions on which a priest will be called in, usually in conjunction with other treatment methods. The priest's power is not limited to any particular illness, but may be applied to any serious or frightening condition that is not responding to other care. The priest's prayers, his reading of the exorcisms for specific ailments, his rituals with holy water, incense, and olive oil, and Holy Communion may all be used to the benefit of the supplicant.

From the interviews with the villagers we learned whom they designate as their healers and what skills they attribute to these people. Questions addressed to a subsample of the community (26 households in Dhadhi, 14 in Panorio, and all 8 Saracatzani households) revealed that the majority of Dhadhi and Panorio families have a family member who possesses special healing skills. Most often that person is the mother or the grandmother in the home. She is not a full-time specialist in any sense; rather, the lore of healing is part of her nurturing role as mother and as the carrier of old traditions, many of which are magical or practical techniques associated with diagnosis and healing.

These are often closely related to the world of nature, which she knows so well: the herbs of the mountains and fields, the birds whose feathers have healing powers, the animals whose bodies, properly prepared, give strength or cure, the stones or amulets that have power to ward off danger, and the ancient words that cast or break spells.

Another aspect of the lore the old women carry is religious and supernatural; they will intercede with the saints and make sure that the ikons of St. Kosmas and St. Damianos, the healing ones who take no pay, hang in the bedroom over the pillow of the sick person; they will call the priest to sprinkle the house with holy water on the first of the month; they will visit the church and pin the silver ex-voto on the ikon of St. Paraskevi (St. Friday), who is "good for the eyes," or St. Modhistos, who "helps the animals." It is the old woman in the household who is most likely to make the pilgrimage to Tenos to pray to the Panaghia for the recovery of her child; and she is the one who may take a tiny scraping from the old ikon on the wall of the little church of the Panaghia in Panorio to bring home to mix in a healing tea.

But the religious lore of the old woman is not limited to Christianity. She knows the habits and abodes of the local spirits and will take pains to warn her family not to expose themselves to the "bad hour," which is abroad at high noon. She knows where the neraides live and of the crossroads where Stringlos the demon was last seen. She knows how to avoid the stars that cast dangerous spells on the mother who has recently borne a child. She knows the danger of sorcery from the priests, witches, and even close neighbors; she will be sure to keep her family aware of the words and deeds that best ward off the jealousy of the gods and the envy of one's friends. The woman, healer of the house, is wise in nature's secrets. She is conversant with spirits and the intercessor with the saints, reluctant before God and careful not to weaken him with her powerful pollution, but less reluctant before the Panaghia, for the All-Holy One is a woman too, and the two can talk of things that only women understand.

Among the shepherd families most households do not have their own healer, but because all the shepherds are kin, it is enough that there are a few wise women in the community. Their best healer is now 83 years old, nearly blind and very infirm. Some still go to her, but she is uneasy about treatment now because she knows her powers have failed. There are two other families in which the grandmother knows the healing ways, but one of these still sends the serious cases to the failing old lady, her peer, whom she respects even now. As for the other grandmother, many shepherds go to her—except members of her own family. Interestingly enough, she is forbidden, by rules she learned long ago, to try to heal anyone in her immediate family. While this stricture reminds one of the Western practice of doctors' not treating their own families, it is more likely to be a hereditary prescription aimed at avoiding danger when a magical operator loses control of the magical power. One can speculate that long ago those wise in the dynamics of family interaction may have wanted to control the power of the old woman and thereby they evolved the prohibition. If this were indeed the case, the present practice may be a survival of that stressful socio-religious revolution in antiquity when the matriarchy was dethroned. It is perhaps significant that the taboo against healing members of one's own immediate family is a tradition that exists only among the Saracatzani. In the other communities studied this is not the case.

Although most of the family healers are women, a few are men. In Dhadhi there are several men with healing skills who specialize primarily in certain disorders that "doctors do not know." Thus one man, who received his knowledge from his grandfather, specializes in curing the wandering navel. Another, who also inherited his knowledge from the male elders of his family, cures both the wandering navel and "the waist." A third cures the wandering navel, and, alone among the men, knows the magical cures for the evil eye. Clearly, the male healers specialize in cures involving hand manipulation of the body rather than word magic, which is often used by women.

The healing skill of the komboiannites is passed on from grandfather, father, or uncle to a male child, although some aspects of hand practika—the ability to "cut the jaundice," for example—can be passed on to a daughter. Magical skills that involve the ability to command the supernaturals, such as the neraides, seem also to be transmitted along the male family line. We infer an exception, however, in the case of some sorcery techniques, which mothers pass on to their daughters; the magical use of menstrual blood seems to be one of these.

The wise woman of the family is likely to have a variety of healing skills. Some of these, such as empirical remedies or the use of herbs, she has acquired from her mother or grandmother or even from older women outside the family. On the other hand, some of her skills have been learned from the older men of her family, and for some of these, tradition requires that she, in turn, pass the wisdom on to a young male of the family. One male healer told us that he had been taught the remedy for the wandering navel by his grandmother, and that he must pass this knowledge on to a female descendant. Should he not do so, the magic would lose its power. This alternate-generation transmission appears to apply in some cases, but not in all, to the words of the xemetrima, to bewitching the moon, and to the cure for wandering navel.

These healing powers, whatever the transmission over generations, may be learned through observation and learning or through "stealing" (surreptitious listening). Stealing is practiced for learning the various words of the xemetrima as well as for learning incantations used to control the supernaturals. The healing powers that are acquired through stealing cannot be acquired in any other way. Stealing may be encouraged, as when an older healer who wishes to abandon his work and pass it on to a younger kinsman purposely speaks the words loudly so the kinsman will overhear and learn them. Or it may occur as a genuine theft when the younger person—usually, but apparently not always, a kinsman—surreptitiously listens and learns the words. There is an advantage to this kind of theft: If the words have been stolen in this manner, the older person does not lose his power, and both he and the thief can practice magic simultaneously. The notion of the transfer of magical healing power as an element or entity, rather than a mere technique, is also found in the belief that a young healer may not be effective until the person from whom he has learned the art has died. The idea that power is directly associated with age is seen in the related but distinct notion that as people grow very old their healing power is enfeebled, and that, as a corollary, the younger ones whom they have taught experience a gradual increase in their healing capabilities.

The majority of respondents in each community said that there are a few people whose healing skills are so well known and respected that they have become "village healers" and are, in that sense, more specialized. These individuals are sought out by others for diagnosis and cure according to the family's tentative identification of the condition and the diagnostic and healing specialty of the village healer. In addition, the decision to see a particular healer in the village is influenced by patterns of friendship or enmity and by the effect of the visit on the balance of obligations between the healer family and the patient family.

In Dhadhi especially, since the village is so scattered, the choice is also determined by physical distance and by the special group memberships of healers and patients. For example, at one end of the village are a group of Attica folk who "stick to themselves" and have little to do with the predominant refugee population of the village. In Dhadhi, also, more families deny there is a special village healer, contending that healing knowledge is shared by all families—especially by the older women—so that it is only one of many skills possessed by family members. "Jack-of-all-trades" skills are necessary to maintain life in a community that has only recently developed part-time vocational specialists and where, for the most part, each family must be self-reliant and self-sustaining.

In Dhadhi the villagers named ten different healers; five of these were mentioned by only one family, whereas one was named by eight families (about 30 per cent of the subsample). The healing specialties attributed to three of the male healers were mentioned above. The fourth man healed only animals. Among the female healers most were credited with knowing the xemetrima for the evil eye, spells, and other ills induced by witchcraft or sorcery. In addition, one woman cured teeth (using both magic and practika), four could cure the wandering navel; and one was famed for her special remedy of mouse oil, which she applied to all wounds. The latter woman was also well known for her wisdom with magic and practika.

Although these specific skills were mentioned as the primary source of the village healers' reputations, each is nevertheless likely to branch out into other healing arts as well. There is some agreement among village families about the variety of skills possessed by the healers, although nothing approaching consensus on roles or skills or persons is ever achieved. This lack of agreement about which people do what is characteristic of nearly all the information we obtained from the peasants and shepherds, and suggests that no matter how homogenous or close-knit a community or encampment may be, there are very great individual differences in how the social world is seen. Perhaps this is one of the many expressions of the individuality so highly prized by the Greeks.

In summarizing the use made of the healing skills of the wise women and komboiannites in Dhadhi, one finds that about half the village families seek out healers to cure the wandering navel, set bones or fractures in men or animals, treat muscular or joint pains, or apply the special cures for the waist. Half the villagers also attribute to their healers—mostly female healers—the power to use magic, the xemetrima and various magical rituals and devices to break spells. Several healers are said to have the power to cure both animals and humans. Only one of the wise women is said also to be a witch; that is, in addition to curing through magic, she is also able to curse, bind, or cast spells and thereby to bring illness, trouble, barrenness, impotence, and perhaps even death.

In contrast to the people in Dhadhi, the people of Panorio are in nearly complete agreement on the presence of village healers with skills greater than the women of the ordinary household. The villagers list three men and nine women. Each of the men was mentioned by only one family, and then merely with passing reference to his ability to cure boils, prescribe herbs, or treat animals. Among the women, five were credited with knowing the xemetrima for breaking spells and curing the ravages of the evil eye. One of them is able to cure painful teeth.

Among the four women more widely known in the village for their healing skills there is one whom everyone agrees is a remarkable healer. She is truly the village wise woman and is so regarded by everyone—even by those who live outside Panorio. Her skills are varied, for she can break spells cast by humans, spirits, or the stars; diagnose and cure ailments caused by the evil eye; and heal humans and animals. Nearly everyone in the village relies on her ability and healing wisdom. We should also note that of the remaining three women recognized for their healing by a number of families in the village, there is one who mixes the bad with the good, for she casts spells as well as breaks them.

Besides the family members with healing skills and one or more part-time community healing specialists whose skills and reputations vary, there are folkhealers who live outside the community and whose names and skills are known to nearly all the local people. Most often these "outsiders" have reputations that are extensive, covering the region or even the entire nation. They are likely to be full-time specialists and, as such (and unlike the local healers), they charge for their services and enter into a more ritualistic or formal relationship with the persons who seek their help.

The people of Panorio, especially, know the famous healer of Spathi, a woman of great power and skill who had the ability to set broken bones, break spells, "cut the jaundice," heal the various ills that doctors do not know, and cast magic spells limited to "the good" (for example, "binding" the jackals so that they cannot attack the lost sheep in the mountains).

It is also among the people of Panorio that Mantheos, the great magician of Spathi, is held in greatest awe. He is an old man now, a true wizard, son of an even greater wizard, uncle of a sorcerer, kin to other magicians of the mountains and the town of Chalkis, said to be rich in warlocks. Mantheos has power over the spirits of wood and wilderness, of night and water. He can gather them at midnight and learn from them why they have struck a villager with paralysis or muteness; he will beat or upbraid the exotika—controlling them in any event—and require them to withdraw or recapture the blow or ill they have sent a hapless shepherd or peasant.

Among the people of Panorio, too, one hears of the centers of magic, Chalkis and Thebes, where mysterious powers abound, strange things occur, spells and enchantment work their wonders for good or ill, and a whole body of magicians—their names not so well known—may be found to cure the incurable ills or break the unbreakable binding curses. In Panorio one also hears of the woman whom they are loath to name, from nearby Doxario. She is evil, so dangerous in her eyes and wishes, that it is better not to name her: and if she is named, her name is spoken hesitantly. She is no healer; she brings pain and sorrow.

Widely known in all three of the communities is the most famous healer in Greece, Vlachos, or as some say, Vlachos the god—a man better known and more admired among peasants and shepherds than any physician. Vlachos is the master of the komboiannites. He is sought out in his Athens office by rural and city people alike for the setting of fractures, and the treatment of sprains, strains, and various muscular or joint discomforts and ailments. Although his treatments presumably are limited to what the doctor would term orthopedic problems, his fame and power are so regarded that persons with other ailments, ranging from blindness to incurable disease, seek him out. Among those who go to him are people from the region of Doxario.

In the town of Doxario itself are several healing specialists known to people in the region. The best known of these is Kostas, who is a komboiannitis but can "cut the jaundice" as well. He is also acquainted with herbs and other remedies. However, he is not considered to be a healer of the caliber of Vlachos. At one time both men were charged with practicing without a medical license; Vlachos won his case and with it added renown; Kostas spent a very short time in jail and upon his release was required to curtail his healing practices. He still does healing, but must restrict himself in order to stay within the law.

The names of other healers in Doxario came up in conversation. There is one woman who excels in bewitching the moon and who thereby may cure the moon spells cast upon mothers, infants, or children, saving them from dread disease or even death. Another woman cures the mumps or, like some others, has special methods with which to treat those skin conditions that the doctors do not realize are illnesses.

Another magician lives in Doxario, a man whose name (Dionysios) was never mentioned by those who believe in his powers. When he was young, it is said, his uncle informed him that he had great powers, and that he must be careful to use them only for the good. He himself says that he has followed the admonition; he knows when to gather the herbs, when the moon is right; knowing what words to say at gathering and giving, he knows he has the power to cure. He may know other, more fatal words too; he may even have forbidden congress with the spirits, although this he denies. But others say differently. He is the one, they say, who has killed with sorcery; angry with a father who denied him gifts under the blackmail threat of curses, he carried out his threat and destroyed an entire family—parents, children, and nearest kin. He is, they say, different from other men because he enjoys women's work; he kneads the dough and bakes the bread; he may even be seen at the spindle or the loom. His voice, they say, is odd and high, and the sound of his light step is sinister. Among those who know him for what he is said to be, there is nothing but dread at the recollection of his deeds. In spite of what the man himself may say about using his gifts for the good, they believe he calls the powers for death. Although the local folk who "know" make it a point to greet him kindly should he cross their path, one can be sure that some will walk a wide mile to avoid him.

THE PRIESTS

An occasional villager refers to the priests as healers. No one reported having called in a priest for illness, although subsequent interviews revealed that some had gone to a priest, shrine, or church in an effort to gain a cure. When asked directly about such visits during the preceding 12 months, two persons in Dhadhi, four in Panorio, and one of the Saracatzani said that they made such visits.

These meager figures do not, of course, reflect the more general use of priests in activities that are primarily religious, but that have a strong health emphasis. For example, the priests come to the houses of those who ask (and pay) on the first of each month for a blessing. The blessing serves to place the house, its family, animals, crops, and fortunes, under the protection of the church, the saints, and God. It protects against any misfortune, particularly against disease, injury, and death. Aside from that, the villager goes to church occasionally, usually on Good Friday, Easter, and Christmas, and for special occasions such as baptism, marriage, and funerals.

The villagers forgot to report their traditional pilgrimages to sacred places. For example, in August the people of Panorio walk the hot mountain paths to the shrine of Aghia Marina, a tiny church near the sea. There they pray, swim, and have a picnic with wine, dancing, and laughter; they drink from the health-giving springs that flow out of the rocks near the church. A priest officiates; he is specially paid for his work. This is, in every sense, a community activity to invoke the good. In addition, individuals ask for the healing of their chronic ailments and kiss the old ikon in the chapel, asking the saint to protect them from further distress.

The people of Dhadhi also go to the shrine of Aghia Marina. Theirs is a different shrine, but their road to it is equally hot and dusty, and the rites of religion and pleasure on a hot summer's day—many stay overnight by the shrine—are the same as those of the Panorio people. Prayers for health, either to be achieved or to be maintained, are an integral part of the ceremony.

Another pilgrimage that people failed to report as a health-seeking activity, but which has a health emphasis, is the trip made to the famous shrine of the Panaghia on the island of Tenos on the fifteenth of August, the anniversary of the sleeping (death) of Mary. During the summer of our study ten people from Panorio, several of whom were quite ill, made the trip to pray for health and other blessings. Four from Dhadhi also made the trip, but none of the Saracatzani went. The rites of Tenos are supervised by priests, and although their intercession is minimal—the pilgrims offer their prayers, ex-votos, and kisses to the ikon directly—the rites are a facet of the healing role of the priest.

In addition to the family festivals and the village-wide celebrations of saints' days and religious observances, there are other occasions when the community is mobilized in religious rituals under priestly guidance. Such gatherings serve to concentrate the combined intent of citizens and priests in an effort to invoke the saints or God for the good of the village. Although no such rituals were conducted during the study period, villagers in Panorio spoke of an occasion, sometime in the 1880's, when an attack by the "bad giant" brought what may have been bubonic plague and threatened the entire population. On that occasion the priests and the community assembled for a number of religious ceremonies and magical rituals, including the plowing of a magic circle around the settlement. These acts were sufficient to drive the giant from the town and to invoke St. George and St. Demetrius, who appeared and slew the intruder with their swords.

There is reason to believe that the reliance on priests as healers has decreased during the last generation. Although they are still called in at death, they may no longer be invited to visit the seriously but not terminally ill. The decreased use of priests as healers is no doubt associated with the availability of physicians and the opportunity to use the hospitals and clinics of Athens. There is also reason to believe that the villagers rely less on the priests for other kinds of comfort and service, and that the priests and the laymen are growing apart.

There seem to be several reasons for the distance that is apparently growing between priest and peasant. None of the three communities has its own priest. The three regional priests are supposed to provide services, but their duties in Doxario seem to prevent them from visiting Panorio more than once a month, and they may not visit Dhadhi at all. Economics is also involved, for the priests (who complain about their very low salaries from the state) expect to receive a gratuity for undertaking the difficult trip to the villages to offer services there. The villagers are unwilling to provide these extra fees because they are already taxed by the government to provide salaries for the priests. Since they have no full-time priest, in spite of their paying taxes, they resent the priests' demand for extra pay.

As Thomas and Znaniecki pointed out in their monumental study of

Polish peasants (66), the holding of divine services maintains the moral unity of the peasant group; the consciousness of a unity of values within a religious framework is reinforced by weekly gatherings; and it is kept alive through the ceremonies that have great social and moral significance. Without these meetings the sense of solidarity in the community may disintegrate, and its feeling of closeness under the guidance of the priest may fall away. This is especially true in a Catholic framework, where the priest is a necessary intermediary between man and God; without the priest's presence, as one woman in Dhadhi put it, "We are not individually so conscious of God; we rely on the priests to lead us religiously. We are not like others [Protestants], who do not need a priest."

In these communities one finds frequent expressions of distrust of the priests and disappointment in their conduct. While these sentiments may arise from unfortunate experiences with one priest, the villagers generalize their antagonism so that all priests must suffer from the bad reputation of the few. The villagers in Dhadhi, especially, report disappointment in the priests; nearly one-fourth of the families in this village think the priests are corrupt and condemn them as "merchants" and "criminals." Specific failings for which the priests, as a class, are blamed, and for which poignant illustrations are given, include rape, murder, bribe-taking, promiscuity, theft, drunkenness, hypocrisy, failure to observe the religious rituals, hurrying the services, revealing secrets of the confessional, viciousness, abortion, and sorcery.

Although there is no evidence that such accusations are true, the villagers consider the failings as genuine; consequently, they show lack of confidence in their priests, even if these are dedicated and holy men. One result of the villagers' attitude is that the sick may not enjoy the comfort a trusted priest might offer. The fact that the sick do want such comfort is strongly suggested, for, despite their disappointments, the shepherds and peasants are very religious. They are careful to fast and to observe the rituals, keeping ikons in their homes and, specifically, placing ikons of the healing saints, Kosmas and Damianos, in the rooms of the sick. Moreover, incense is burned in the rooms of the sick, prayers are said, and pilgrimages are undertaken to various shrines and holy places.

Although some of the layman's distrust of priests may be attributed to simple economics and perhaps to dissatisfaction over the shortage of priests in the region, there are other contributing factors. Priests are men of magic, and as such they are sacred. The powers with which they deal are great, and the peasant stands in awe and dread of them. This very power and sacredness of priests creates an uneasiness, and when any untoward circumstance does arise, the priest's role is likely to be exaggerated and his motives misattributed.

In this regard Thomas and Znaniecki have observed that among Polish peasants the magico-religious system produced many negative effects. Insofar as the priests have taught the peasants over the centuries that there is magical causality, that God has magical power over things, that this divine power can be directed by the intentions of the priests in harmful or beneficial ways without any necessary moral basis, the peasant believes that anything can happen. He lacks, say Thomas and Znaniecki, a sense of rational cause and effect because his intellectual superiors, the priests—and the magicians before them—have educated him to be credulous. This expectation that the intention of powerful magical operators—or of the spirits themselves—can intervene to disrupt ordinary natural processes, leads to a sense of helplessness among those who lack the power to control magical forces or energies. The peasant is weak in this regard, but may believe his priests or magicians to have an almost unlimited control over events. As long as the peasant subscribes to the magico-religious system, he has no consciousness of the limitations of power among his superiors, and must ascribe to someone the responsibility for anything significant and unusual that occurs. Since the peasant deals from a position of weakness, his only weapon is "cunning, apparent resignation to everything, universal mistrust, [and the wish to derive] all the benefit possible from any fact or person that happens to come under his control" (66).

In such a state of affairs, one can see how the peasant tries to manipulate and influence the priest without feeling any loyalty for or trust in him. Should things go badly, or should the priest appear unresponsive to the peasant's overtures, the peasant blames the priest for his negligence or for his evil intentions. On the other hand, if, under the influence of his contact with city folk, the peasant learns to be skeptical of the whole magico-religious system, he has no reason to need the priest at all, at least until the emphasis of the church shifts from a promise of magical rewards to a concern with moral and spiritual matters. This latter development, which characterizes urban Christianity, is not particularly apparent among rural churchmen, who have themselves been reared in a system that is more magical than spiritual in emphasis. Some of the unfortunate national scandals involving the hierarchy of the church in recent years are known to the country folk, and these augment distrust.

Traditionally, the Greeks have been independent thinkers and skeptics. Although they are proud of the role of the church in the fight against the Turks, they are also skeptical of some of the political and economic activities that they think contemporary churchmen engage in. Perhaps this too accounts for our finding that a strong minority of families, especially those in the more urbanized village of Dhadhi, indicates that the need for priests is past, and that they want nothing further to do with them.

Yet in spite of what villagers say, there is still a bond between peasant and priest. The church remains an important influence in national and local life, and religious sentiments pervade every home. In times of sickness it is a rare person who does not attempt to purify himself, whose family does not pray to its chosen saint, or in which one of the older women does not consider a pilgrimage (if the illness is serious) to a healing shrine.

12

Renowned Folkhealers

In this chapter we shall discuss the most prominent folkhealers in the region, healers who are most often sought out because they have knowledge and skill surpassing that of the old mother in the family or the men and women in the neighborhood. We shall describe the viewpoints and some of the personal development and characteristics of the wise woman of Panorio, the komboiannitis of Doxario, the great magician of Spathi, the evil magician of Doxario, and the famous komboiannitis of Athens.

The folkhealers share the culture of the peasants and shepherds; they are a part of this group regardless of their reputations and specializations. They have the same assumptions about man's relation to man, nature, and the supernaturals that guide the lives of the uneducated people of the region. It is true that some of the healers have built a superstructure on this heritage and now possess an elaborate body of knowledge and skills, but the superstructure differs only in detail from the peasant and shepherd view. Thus the great komboiannitis of Athens, "Vlachos the god," as they call him, now reads x-rays and has a table full of anesthetics in his anteroom; but even though he may use some of the techniques of medicine, he is not technically oriented. The great magician of Spathi knows much about the supernaturals and how to invoke the neraides or banish the witches; but his skills are based on the same methods of magical control that are practiced in a limited form, by the family grandmother when she says the xemetrima or recites a curse at the crossroads to ward off the impending attack by a demon. Only the priests have moved markedly beyond the traditional culture, inheriting from the innovators and formulators who preceded them a grander edifice of belief. Some of the city priests especially are religious rather than magical men, and are imbued, as Redfield says, with the "great traditions" (56). That progression—from magic to religion—is a historical and psycho-social process well described by Paul Radin in *Primitive Religion* (54). But in the country the priests are involved in magic as well as religion; the country priests, coming from peasant backgrounds, allow the little traditions (i.e., country ways) to exist side by side with the great traditions. It is this very accommodation that binds them to the rural folk.

MARIA, A WISE WOMAN

Among the people in the three communities are, as we have indicated, many wise women and a number of men with limited healing skills. But among them there are only two who are known to nearly everyone in their community as particularly powerful healers. These two are the old grandmother of the Saracatzani camp, now blind and failing, and a middle-aged woman in Panorio, a bright, sensitive, gentle person who is rich in healing lore and helpful skills. The old shepherd lady carries her 83 years with sad infirmity and could not easily be interviewed. But Maria, the wise woman of Panorio, was most communicative; members of our team saw her nearly every day and enjoyed the warmth of both her own and her family's friendship.

While there is much about Maria that deserves reporting, we will begin with the fact that her position as the village wise woman does not set her apart from others in her village, for she is one of them in dress and belief. Although her special skills are acknowledged, she receives no homage, nor is she feared. Her approach to illness is essentially helpful; she wants to do what she can for the welfare of others. She charges no fee and demands no gifts. Insofar as she has a long-term goal, her work is for the good of her soul. She does not compete with the physicians or magicians; while the richness of her stories and the delicacy of her feeling show that she is a dramatic person, she performs no dramas to charm or hoodwink others. If her secret words, her incense and flowers, do not aid the sufferer, she is the first to suggest that he seek more skilled care at the hands of another. Interestingly enough, she does not recommend that they go to a folkhealer but to the physician. In doing this, she probably persuades people to seek medical advice who would not ordinarily go to a physican, if left to their own devices or to less modest healers.

Maria and her family use physicians, and have averaged more visits to them the past year than the villagers as a whole. She has high regard for medical science, and appreciates what physicians did to save the life of her daughter who had been ill recently. She views her own healing methods with a certain amused objectivity, saying that if, for example, her husband has the flu, she will try sage tea and warm drinks, and perhaps a little magic, but if she cannot cure him with "the fairy tales," then "we run to the doctor." In her mind the city ideas coexist with country beliefs; she knows both the great and the little traditions, and she is capable of viewing either with humor, intelligence, and a pragmatic, skeptical open-mindedness. It should be noted that her remark about the fairy tales was not made defensively against the possibility that the charge of "superstition" would be leveled at her by the "city people" of our team. The city people in Greece that she has known—and that we have known—are often as well versed in the lore of folkhealing as their rural relatives. Maria knows that her traditions, some of them dating back to classical and pre-classical Greece, are "fairy tales" in the sense that they are ancient, but they are part of her world and she believes them. Like most Greeks, she is not committed to any one system to the exclusion of all others; she will use science and religion and magic simultaneously if need be, and will not be troubled by any sense of inconsistency.

She and her family attribute the development of her healing powers to her own illnesses and suffering, her earlier vulnerability to the evil eye, and her successful struggle to overcome her pain and distress. When ill herself, she learned cures from older women; one of her teachers, for example, was the famous midwife-healer of Spathi (see Chapter 11). Maria also learned from elders in her own family. Her grandfather was a powerful magician but an evil one, a sorcerer who bound his own son. Maria's mother had the opportunity to learn the magic arts from his hands, but she is said to have refused. Nevertheless, we must suspect that some of the old skills have been handed down. Maria's dedication to working only good with magic probably reflects values she learned from her mother, who was highly critical of the grandfather's sorcery. Maria has lost three children whom she loved; her lovely daughter nearly died this last year, and one younger child is painfully afflicted. She knows sorrow and bitterness, and is dedicated to helping the living so that they may not die before their time. She is very much concerned with the health of the infants in the village, saying that she tries especially hard to make them well if they are sick, as if, in a way, she is protecting others from the loss and mourning she has suffered. "I do not want anyone else to taste such bitterness," she says.

If someone comes at three in the morning, she will awaken without complaint to try to help, whether the patient is an old man or a sick goat. It is satisfying for her to be able to give aid and comfort, and her family is exceedingly proud of her devotion and reputation. Her gentleness and generosity are unusual in Panorio, which is rent with hatred and selfishness; not many would act as she does. The very unusualness of her actions and concern for others indicates the intensity of her motives—motives that reflect a genuine love for others, an attempt to overcome her own grief and pain and a desire for the esteem of her family and neighbors. Her healing efforts transcend the quarrels and factionalism in Panorio, as the following story told by Maria's daughter shows:

One of the Petridis' baby boys was sick. He was only a few days old and not yet baptized. The parents brought the priest to baptize him so that he would die a Christian. They called in Mrs. Plastira [another wise woman in the village] but didn't call Maria because Maria and the Petridis had had an argument, and the Petridis were afraid Maria would not agree to cure the baby. Since nothing helped the baby, neither the priest nor Mrs. Plastira, one of the relatives who knew about the fight and who was a friend of Maria's, went to her to ask a favor, not telling her what it was until Maria agreed she would do whatever was desired. Only then did the Petridis' relative tell her about the baby's condition. Maria felt very bad because the Petridis had believed that she would refuse to treat a person, especially a baby. Maria hurried to their house but did not enter in order that they would not feel she was going to harm them in any way. The Petridis brought the baby outside. Maria had to say "the words" three times before the baby started getting better, but the child did recover.

Maria's actions in this case were not only magnanimous and "good for her soul" but consistent with her moral obligation to heal.

Maria abides by her creed not to take money, although she may take a gift, as the following case illustrates:

Once a woman from Athens was visiting in the region. She had an eye difficulty; the "white spot" had developed in it. Someone saw this and told her that Maria could cure it. She came immediately to Maria even though it was night. Although it is the belief that "the words" should be said when it is daylight, Maria decided to attempt the cure. She did not give her anything to drink [herb infusions or special preparations] for she never does [give medication or preparations to patients]. She merely said "the words." She was successful and the woman wanted to pay her, saying that she had spent a lot of money going to doctors and that they had done nothing; now that she had been healed, she wished to pay. Maria did not accept. Later the woman came with some coffee beans, which she offered to Maria. Maria felt that she could not reject her wish to thank her, and so she did accept the coffee beans.

It will be useful, in trying to give a clear picture of the operation of folkhealing and some of the beliefs surrounding it, to present Maria's comments on one condition, the evil eye, in some detail:

When a person is bewitched, he shivers, has headaches, is nauseated, sees black, and may have to stay in bed. Sometimes the shivers are as bad as those of people with malaria; they jump two feet out of bed with them. If you are bewitched and do not go to someone to break the spell, your situation may become serious. The effects of the evil eye become "knots," and they move far inside you. When you are first bewitched you have the headache; later your bones are affected and they hurt; then your heart becomes involved, and you may have diarrhea and vomiting. Your stomach becomes like a ball of thread. They say the reason the heart and stomach are both involved in this last stage of pain is that the discomfort follows the line that goes between the heart and the stomach [when the people of Panorio say "heart," they touch their bellies]. When the bewitchment is in this last stage, most of the women who can do the xemetrima do not sense it; they cannot recognize it [the bewitchment by the evil eye] any more.

You diagnose it by saying the words and at the same time putting olive oil in water. If the oil disappears so will the distress, and you know that the person was in fact bewitched and that now [through putting the oil in the water and saying the xemetrima] he is cured. If the oil does not disappear, then it is not the evil eye but something else that is causing the discomfort.

I do not use the olive oil at all times, but I am able to sense, as I say the words, whether the person is bewitched, and what is the extent of the bewitchment. One way I can tell is that I get the discomfort myself; I start to yawn or get a headache; generally the pain comes to me.*

* Barber, in discussing the use of hypnosis as an analgesic, cites the work of Butler with terminal cancer patients. Butler is quoted as saying that the physician must give of himself, that giving hypnotic analgesia tires the hypnotist, and that as the bond between the two grows stronger, the hypnotist-physician may actually feel the symptoms he is trying to eradicate from the patient. (T. X. Barber, "Toward a theory of pain: relief of chronic pain by prefrontal leucotomy, opiates, placebos, and hypnosis," *Psychological Bulletin*, 1959, pages 56 and 430–60, citing B. Butler, "The use of hypnosis in the care of the cancer patient," *Cancer*, 1954, Volume 7, page 6.) The correspondence between Maria and a modern physician is remarkable, and suggests similar psychological processes operating in this kind of healer-patient relationship.

In treatment the words* that are used are both Greek and Arvanitiki [Albanian], some in one language, and some in the other. While I say the words, I either put olive oil in the water [in a cup] or put incense on the coals [in the fireplace] so that the bewitched person is perfumed with the smoke. Sometimes I put flowers on the coals, flowers saved from Easter that were the body of Christ during the holy ceremony of the burial of Christ on Good Friday. The bewitched person is perfumed with the smoke from the flowers. I feel I have great power to break the spell of the evil eye; I don't know why, but I feel it so strongly that if someone were brought to me dead from being bewitched I could give him back his life, so to speak. I am not able to treat myself for the symptoms I get from other people at the time I free them from the spell. I had wondered about this often, and finally discussed it with other [wise] women. They told me that the bewitched are relieved just because I get the pain from them. Should I do the xemetrima on myself, the pain remains with me; I cannot pass it on to someone else. I guess that explains it.

Animals can be bewitched too, and I have special words to use on them. I use the words ordinarily reserved for mules on most animals. These words have to do with the vines of the grape. In very serious cases I may use these on people, too.

Not everyone can be bewitched, and not everybody has the evil eye.† Some people have magnets in their eyes and can bewitch without even knowing it when they admire something. They say that people with eyebrows growing close together also have the evil eye. Most of them do not even know they have this ability. If they do know it, then they should always spit on the person or animal or thing they admire. In this way they counteract their own spell. The people who are more likely to be bewitched are the ones who have developed a habit of becoming victims of the evil eye. These are persons who as children had their parents do the xemetrima on them, and they have become used to it. Even for these people there is one factor that is especially important in their being vulnerable; it is their blood. There is one kind of blood that is easily bewitched.

There are many acts that can bring the evil eye on a person. One, for example, has to do with babies. When you decide to wean a baby, you should follow your decision strictly. If one day you do not give your milk to the child and then the next two or three days you give him your milk again, your baby can be badly bewitched. This means almost death for the child; you should never do it. Nowadays the younger people do not believe this, but the older generation certainly does.‡

* Maria refers to the words of the xemetrima as "lies" and refuses to divulge them. One is reminded of the ancient practice, cited in Harrison (30), of hiding the sacred behind the profane.

† Maria was unwilling to tell us who in the village has the evil eye, but she told us who else can break the spells.

‡ Irritability and distress in the infant are usually attributed to the evil eye: one can see how diet change, especially if it includes poor feeding practices such as giving the infant unboiled whole milk—as is still done—may well produce serious gastrointestinal

Although we will not dwell on the interesting concepts included in Maria's discourse on the evil eye, we must at least call attention to the notion of sickness-transfer, in which the pain is drawn from the body of the bewitched and taken into the body of the healer. The idea here is not far from the ancient notion of the animated spirit of disease, the Ker (30) that moves willfully from person to person and can be manipulated by operators with control over magic. Nowadays the healers will say, in this regard, "May it go to the mountains," as they banish the spirit of illness, wiping it away as Asclepius wiped away diseases (47). The notion of transfer is also intimately associated with the "pharmakos," or remedy, which meant in ancient times cleansing and expiation through purificatory rites of sacrifice. In these rites the infection of pollution—the unclean, damaging power—was wiped away, transferred to the scapegoat (the pharmakos), which was subsequently killed. One member of the community took upon himself the burdens of the others and suffered to save them—as Oedipus and Christ did.

Maria's concepts of the evil eye and its treatment are much more elaborate than those of the other villagers. She is also better informed about hygiene and health matters, more fully acquainted with diseases that are dangerous or contagious, and richly aware of the traditions and folklore of the region. She has superior intelligence and considerable verbal facility. Although she has never gone to school, she is better informed on many matters than most of her neighbors. But in spite of her knowledge and charm, she is very unsure of herself; she is nervous, suggestible, easily made anxious, and easily hurt. She has considerable insight, and discusses her feelings of inadequacy and impotence resulting from the death of her three children whom she tried unsuccessfully to save.

Her respect for the medical profession remains, even though both her male infants, the ones she loved the most, died while they were under the care of physicians in the region. One child died after the physician told her the baby had a cold and need only be rubbed. The other child also died after a physician had diagnosed a cold. In the latter case, the women of the village were in strong disagreement with

---

disorder. In addition, Greek mothers are expected to be firm about weaning; often they put pepper on the nipples. Should a mother not fit this cultural expectation and show herself to be weak or vacillating instead, her internal conflicts or her distress over the inevitable criticism from others may well lead to anxiety that is communicated to the infant.

the physician, insisting that the infant had been struck by the "bad hour."

Maria does not blame the physicians for these deaths; instead, she seems to blame herself. More than that, her confrontation with death continues; she thinks about it often, and, in a fashion common to Greeks, anthropomorphizes death so that it has become a very real person to her—Charon. She tells of conversations with him, and of how she has told him she is willing to go away with him if only he will let her live to see all her daughters married. One senses that denial or easy rationalizations of death, so often seen in others, do not suffice for her, for she is trying to develop a personal philosophy, or at least an orientation, that allows her if not to accept death at least to deal with it openly and deeply. Maria is, in a sense, one of those formulators of religion whom Radin described (54).

Perhaps it is in this sense that we can speak of her as "light-shadowed" (elafroiskioti). This interesting concept will be discussed at length in a forthcoming book, *The Dangerous Hour*. Let us note here that the "light-shadowed" are people described as simple, good, and pure. This quality is associated with second sight, the capacity to sense directly the exotika. All rural folk know of the exotika and accept their existence, but only a few are able to see the neraides and demons, the woodland spirits, and the other supernaturals who have inhabited Greece since ancient times. Politis (53), the Greek folklore scholar, thinks of the light-shadowed ones as similar to the epileptic shamans of more primitive times. These people, according to Radin (54), suffered deeply because of their constitution; their interest in the supernatural world expressed their own psychological conflicts, which they resolved by becoming one with the hidden and mysterious. One sees no evidence that the epileptic in contemporary Greece has any spiritual or shamanistic role, and under no circumstances can Maria be diagnosed as an epileptic. But she has experienced the suffering and shows the internal conflicts, sensitivity, and romantic or mystical bent that Radin describes in the second-sighted ones. Maria is a genuine healer whose personal characteristics are not far removed from the shamans of primitive societies.

We do not know if Maria can actually heal the sick, for we have not made therapeutic results the concern of our study. Maria believes she can cure, and everyone in her village believes she can cure. They all cite cases to prove their point. They are probably right. Jerome Frank, in *Healing and Persuasion* (24), reviews the results of magical and religious healing and concludes that benefits do accrue. He mentions the following conditions for the magical cure of illness: The healing efforts must arouse the hope of the patient for cure from a healer who mediates between the patient and the group. The healer must have an institutionalized role with powers attributed to him by the group. The healer and patient must share with the society common assumptions about illness cause and necessary healing ritual. Finally, healer, family, and community must be involved in mutually supporting service, and intense emotions must be invoked. These conditions hold for the work Maria does in Panorio, and for this reason we assume that she is at least occasionally successful with somatic illnesses. Psychosomatic illness as well as illnesses caused by interpersonal stress certainly may very well yield to the persuasive force of the shared expectations that Maria administers.*

KOSTAS THE TURK: A KOMBOIANNITIS

Kostas the Turk, so-called because of his origins in Asia Minor, lives in the central community of Doxario. Kostas is well known throughout the region for his skill as a komboiannitis; he is said to be able to cure headaches, the wandering navel, sprains, and fractures, and to cut the jaundice as well. He lives in a typical tile-roofed adobe hut in Doxario and works as a peasant tilling the soil. In days past he was much busier with his healing work, but, as mentioned earlier, one of the local physicians filed a complaint against him a few years ago, accusing him of practicing medicine without a license (he had set a broken leg). Kostas spent several hours in jail until a senior police official came—so Kostas tells the story—and released him because he had cured the official's wife. Nevertheless, both his practice and his prestige have suffered as a result of the incident.

There is considerable animosity between Kostas and one of the two local physicians. Kostas blames this doctor—who is not a native of the village—for his persecution. The other physician in Doxario, an older

* The confidence the villagers have in her, and her genuine dedication to the alleviation of pain, suggest that Maria and women like her elsewhere are a potential health resource. In the absence of sufficient numbers of nurses, health educators, or rural doctors, one can conceive of rural centers' training women like Maria for practical nursing roles in the villages.

man who grew up in the village, seems to have a better relationship with Kostas; at least Kostas does not include him among his persecutors. In contrast to his feelings toward the physician who brought charges against him, Kostas cites as an "excellent man of rare character, a true physician," the old doctor Kefalas, who lived in Doxario until his death a few years ago. Dr. Kefalas "accepted" that the wandering navel was a disease and referred such patients to Kostas for treatment. It appears that Dr. Kefalas referred other patients to Kostas too, so that the working relationship between the physician and the folkhealer was very close.* Dr. Kefalas shared and respected the traditional rural beliefs about illness; the younger physicians of today do not.

The younger physician from out of town, representing the growing influences of the city, technology, and "modern" ideas, has on his side the laws made by the educated legislators in Athens. Kostas sees these changes and resents them. He is the victim of technological change; his stature in both the community and the region is reduced; his healing work, which was a source of pride and satisfaction to him, is greatly restricted, and his life is made more difficult because he is losing in his competition with the younger physician. He says:

If I were educated, no doctor could compete with me. But now I'm persecuted; I know how to cure the navel and broken limbs, as well as how to use special herbs that would bring me much money if I were free to use them. The doctors do not know how to cure the jaundice and the navel; these are not even recognized by science. The doctor doesn't know what to do with fractures or sprains either, unless by chance he is a gynecologist. Gynecologists have to know about fractures because it is at delivery that most such accidents happen.

The doctor relies on his science, but the practikos acts according to experience and practice; also people approach us differently. When they are with a doctor, they must show respect and when they speak they must be careful about their vocabulary and accent. When they come to a practikos they can be more casual. Many times we treat our patients to dinner. Besides that I do not ask for pay. I just try to do my best, and I never accept money or gifts until the patient sees for himself that he is feeling better. I am not like the doctors who have a price from the very beginning. The patient does not have to pay me—certainly not if he doesn't get better. In any case, it is up to the mood of the patient whether he pays me or not.

* The practice of physicians' making referrals to folkhealers is known in many places. In Italy physicians may send their patients to magicians, and in Puerto Rico psychiatrists and mediums sometimes share patients (57).

We see here some very important differences between the more formal folk-healing practice of Kostas in contrast to Maria, for example, and the full-time specialty of medicine as it involves healer-patient relations. With Kostas, payment is only for cure, and then on the patient's terms. There is no business arrangement or contract; the relation is an extension of the informal mutual-help tradition of the peasants, reinforced, as that practice always is, by the obligation of the receiving person to pay back his debt. The mutuality and obligations here need not be defined by fee or price or agreement, for both Kostas and his patients are fully obligated to one another by the demands of the culture they share. Healer and patient naturally understand one another in this setting; they need put on no special façade. Kostas can have his patients as dinner guests because they are status equals; his patients need not mind their clothes or manners, because their healer is a peasant like themselves.

Kostas, like other healers, has learned his art from his family and from skilled older persons. He says of his becoming a healer:

> In my home town in Asia Minor the practikos was the only doctor. As a person everyone respected him; he had a high position. This is not my case today, for the people here [in Greece] are not as kind and as hospitable as the Asia Minor refugees; people here are bad and want to harm one another. Well, in any event, my uncle in Asia Minor was the practikos doctor in my town. He knew how to cure the fractures and the sprains. Everybody, including the women, was interested in my uncle's skills, and we all learned by watching him work. I loved to watch my uncle. Gradually I learned healing, but it was not until I was 24 years old that I did it on my own for the first time. Actually, no one knew of my capacity as a healer until one day my prospective bride's mother broke her leg while she was working in the field. There was no one there to care for her, so I did, and I cured her. After that I began to become well known here in the Doxario region. That's when I changed my name to Kostas because the name I had before was a pure refugee name, and a policeman friend told me to change it so that I would not be falsely accused of crimes. At that time [1922], the people here accused the refugees for all the minor crimes that occurred during the disturbances then. Well, I changed my name but gradually people began to call me the Turk because of my accent.
>
> I should also say that I also learned healing from a Turk who knew how to cure the jaundice and the navel.

Apparent here is the antagonism between the native Greeks and the refugees from Asia Minor who were resettled in Greece in the early

1920's after Greece lost the war with Turkey. That antagonism still exists. One suspects that Kostas finds it easy to fit his present predicament into the pattern of persecution and discrimination, which the refugee people say they have experienced at the hands of the homeland Greeks. It is interesting to note that Kostas learned how to treat the wandering navel and the jaundice from a Turk. Many of the folk beliefs (such as the belief in the evil eye and in revenants) and practices of the rural Greeks are shared with other people of the Near East and the Mediterranean basin.

We asked Kostas how one cures the various ills. He replied:

> When somebody suffers from the jaundice, his color turns yellow and he feels weak. You cut a membrane on the inside of the upper lip and put some salt on top of it. Then you make an incision in the forehead and on the top of the head and on both temples, after having made the sign of the cross on each one of these four places.
>
> When the navel falls, a person feels weak and thirsty. The healers collect the [wandering] navel from various parts of the body. I collect it in the armpit. After I find the special nerve in the armpit, I rub it for some time until it becomes as big as an egg. Then I give the patient a hot drink and ask him to eat good food for some time until the symptoms, which are weakness, vomiting, thirst, and dizziness, disappear. I also tell the person not to carry heavy loads on his back until he has regained his powers.
>
> For a fracture you wash that part and feel the broken part. You put it back together and then prepare a plaster by beating together the white of an egg, some soap, and ouzo until it is an ointment. You put that around the broken part.* The more recent the fracture, the less painful it is. The plaster will get loose by itself when the broken part is cured. It takes as many days for the fracture to mend as are the years of the patient.

Each treatment method is relatively simple. For jaundice the ritual includes religious symbolism and the specific cleansing of the body of the illness through blood letting (which occurs at the time of the cutting). As Kemp (37) has observed, the folk-medical practices of peasants usually include a mixture of religious and magical acts, commonsense care, traditional ideas, and medical notions that have been filtering down since the time of Hippocrates. Kostas, however, limits himself to the work of the komboiannitis. He does not have anything to do with the exotika. If the patient's problem is one of sorcery, or of having been

* The practikos soaks a bandage in the plaster and heavily bandages the area.

struck by the exotika, Kostas refers him to Mantheos, the great magician of Spathi. Kostas has no wish to deal with magic, for it is a dangerous business. One slip and the exotika will turn on the magician.

MANTHEOS, THE GREAT MAGICIAN

Higher up in the mountains, a few miles farther on the winding narrow road from Panorio, is the town of Spathi. It is picturesque, with flowering Judas trees shading the public square, about which the tavernas stand. It is a dramatic place; many stories are told of the murders and blood feuds, the sorcery and magic that have occurred there, and of the passionate and intense people who have lived there. There are several well-known healers in Spathi; the most famous is Mantheos, who is descended from a family of magicians—his father was Miltiades, the greatest wizard of all, so they say. Mantheos is 76 years old, a full-time specialist in the sense that he does not till the soil, and, like a doctor, charges for his work. He sees from 60 to 70 clients each month. Unlike the komboiannitis, Mantheos is suspicious of visitors and will not discuss any of his healing methods. Just as some of the wise women guard their magic words, and as Maria refused to tell us her "lies" (although others would recite the incantations), Mantheos jealously guards the means he uses to control the powers. Our visits to him did not take place under auspicious circumstances. On both occasions he was in Athens in the hospital and quite unwillingly there; he was under the care of doctors whom he considered hostile to or skeptical of magic. He even refused to admit that he was a magician. He did imply that a visit to his home when he was well might find him more willing to talk. One of his daughters, whom we visited in Spathi, agreed, and further suggested that $2.00, paid on the line, might make it easier for her father to converse.

From his daughters we learned that Mantheos and his brother both had been taught the secrets of their powers by Miltiades, their father. But the brother did not learn them well and was unable to cure. Mantheos learned them well, and in addition has special powers—the powers of God—so that every illness which is brought to him and which is susceptible to his power, i.e., caused by sorcery or the supernaturals, will be cured. Among these conditions are epilepsy, nosebleeding, impotence, the breast pain of the nursing mother, difficulty in child delivery, the exotika, and the "bad hour."* In his usage the exotika are likely to refer to supernaturals' taking human form, especially a female form, or to supernaturals' taking animal form, especially a male form. Mantheos is careful to evaluate each client's problem immediately, telling him in advance whether he can treat the case.

Mantheos has learned his skills from family and will pass them on to family. He is saddened because his only son lives in Athens and has no interest in this work. The son has abandoned the old ways. There are five daughters living in Spathi, and Mantheos insists that one of them must carry on his work because it is of such great use and help to mankind. However, the daughters are reluctant to learn, although one of them is already considered by the villagers to know enough of the techniques to apply them. She is not willing to admit this to a stranger, but her admiration for her father and her conviction suggest her interest. The only family member who is really interested in learning the techniques is a son-in-law who lives in Doxario. Mantheos refuses to teach him because he is only a son-in-law, not a blood relative, and also because this man took a second wife as soon as he became a widower, although his own daughters were already of marriageable age.†

Mantheos uses no herbs or plasters. He works only with words and rituals. These words, the many forms of xemetrima, are in Arvanitiki. The words need not be understood by either the magician or the sufferer for the cure to take effect. In order to teach his daughters the words, Mantheos deviates from the secretive practices and speaks them loudly so that they can "steal" them.

Asked to describe a successful case, one of his daughters told us the following:

Some years ago a young boy from Euboea had an epileptic seizure in front of Mantheos and his daughter. They asked about it and learned that the doctors had been unable to cure him. Mantheos, on hearing this, took over the treatment. He gave the boy instructions about how to live for 40 days. These included fasting, the avoidance of self-pollution, and the avoidance of contact with any polluted persons. Mantheos told the boy that if he followed his instructions exactly he would be cured. The boy did so and was cured.

* Note that the nomenclature here finds the disease name and the disease cause synonymous; also note that the magician differentiates between the exotika, a general class, and the "bad hour," which for most folk is merely a subclass of the exotika.

† Mantheos seems to feel that his son-in-law should not have remarried, but should have devoted himself to securing good marriages for Mantheos's granddaughters.

[It is likely that Mantheos also said the xemetrima.] Now whenever a villager from Spathi passes through the man's village in Euboea, the man gives the traveler a greeting to Mantheos, expressly saying, "May the incense grow on his grave" [as it is supposed to have done on Christ's grave]. The man says this because Mantheos is, to that patient, a holy man, the Savior, and even now he worships an ikon dedicated to Mantheos.

Mantheos does only good magic, but he knows black magic too; he must know it to be able to counteract sorcery:

There was a rich Greek-American widower who came back to his village in Euboea and got married to a poor, but good and handsome, girl. The people of her village envied her, and someone there cursed them with a binding curse. When the couple went back to America, they could not have intercourse because they had been bound. The man wrote to Mantheos explaining the situation and asking him his advice. Mantheos wrote telling the man to send him a "sign" [a personal belonging that had been used by the man and his wife]. He did so. Mantheos was then able to unbind the sorcery; within a month the woman became pregnant.

DIONYSIOS THE SORCERER

Mantheos is a guarded man, but his work is known to all. No one hesitates to mention his name, and while all would agree that he knows black magic, they also agree he uses it only to fight fire with fire. But in Doxario there is another magician about whom the mystery is deeper. In this case it is not the magician who is unwilling to talk; rather, it is his neighbors who are unwilling to talk about him. His name is Dionysios, and they say he is a sorcerer. He is blamed for many things; when he passes a donkey, the animal becomes crippled; when he passes a tractor, the engine fails; and when he is angered or slighted, whole families may die. These are not apocryphal tales, mere legends, but are told by people living now who blame Dionysios for the deaths of their friends and relatives. They refer to him in various indirect ways, but never once did a person who believed in the sorcery of Dionysios mention his name. He is referred to in terms of his idiosyncrasies, oddities that mark him as a deviant in the village. He is "the one who does woman's work." "He speaks like a woman," they say in reference to his high voice, or "He speaks as my grandfather would, in the manner of the old days," referring to the unusual speech patterns he uses. Everyone, they say, hates and fears him. No one will dare to kill him, but all expect that he will suffer very much when the time comes for him to die, for the soul of a bad man is said to have a hard time leaving the body.

We visited Dionysios when he returned late from working in his fields; it was the full of the moon. His house is on the edge of town, somewhat isolated from the others and protected by a small, mean dog. Dionysios is a small man, about 62, with a sometimes shrill voice and quick bird-like movements. He does not appear sinister, nor does his wife. However, when he begins to speak confidentially, as is his wont when talking of magic, the pitch of his voice, like the thin whistle of a distant nightbird, can be disconcerting.

Dionysios learned his magic as a boy from his second cousin, the brother of Mantheos, son of Miltiades, the man who Mantheos's daughter says did not learn magic well and could not cure. There are two tenets to his work, says Dionysios: never to charge for healing, and never to do evil with his powers. No one, he insisted, had any reason to fear him; but when asked why villagers might fear a magician, any magician, he declined further discussion. When the time comes he will pass his magic on to a son, although he states that there are no prohibitions against a daughter's learning it. It is only a matter of whom the parents want to teach, and of the interests of the child.

He does not consider himself in competition with the physicians, and, like Maria, refers those he cannot treat to them. He distinguishes, as most peasants do, between those diseases the doctors know and can cure and those that only the folkhealers know and can cure. Among the illnesses in the latter group, in which he specializes, are those caused by the moon: the "anemopyroma" (a kind of swelling, perhaps erysipelas, especially of the facial type), the jaundice, the evil eye, and others he refused to name. "My method," he said, "is not science, but it is knowledge and it can be effective. Even science does not have the power over the moon. I can influence the moon."

He cures ailments by using the xemetrima combined with certain plants or herbs, which he collects at special times on the mountains or on the plains. While cutting each plant he speaks an incantation, which includes the Lord's Prayer. When telling us about this, he pointed to a bundle of herbs in the corner of his yard; we asked if we might look at them under the light so we could see them better, but he refused. (We had already been given a collection of 20 or so herbs by various wise women, and another practikos; even the priest gave us two healing herbs that he uses at home.)

Asked to describe a case he had treated, he told us the following:

There was a young man who was under care for a long time in the hospital in Athens. His skin had turned yellow and he continually lost weight. The doctors could not diagnose his case despite the equipment and the laboratories they had at their disposal. His mother finally came to me and gave me the history of his case. I understood that he was affected by the moon. I boiled a special herb and gave the broth to the mother, who took it secretly to give to her son in the hospital. Immediately her son began to recover. The doctors were amazed. They wanted to find out what the juice was that he had drunk and that cured him. The mother denied she had given him anything. The boy was released from the hospital and is now the father of a good family.

Dionysios's case demonstrates again the common belief that jaundice (presumably that was the diagnosis) is not recognized or treatable by doctors, and that successful care requires the intervention of a folkhealer. Their treatments differ; Kostas, the komboiannitis, "cuts" the jaundice; here Dionysios uses magic words and herbs sanctified by ritual gathering. The case above serves to prove to the listener the superiority of folkhealer over physician, and the technique used—sneaking herbs to the hospitalized patient—might well give the folkhealer an opportunity to claim credit for other improvements in hospitalized patients, since recovery can always be attributed to the herbs and xemetrima rather than to the physicians. It is important to note that it is the woman, the mother in the family, who in a nurturing fashion seeks a traditional peasant cure for her child. The folkhealers do not object when two (or even more) forms of treatment are being tried at once.

When asked about treating the illnesses caused by the exotika, those supernaturals over whom his cousin Mantheos has such remarkable powers, Dionysios denied that the exotika exist. He granted that they existed in the past, but said—as do many—that with modernization "we have become the demons ourselves," and the old powers to see the exotika are gone. On the other hand, he added that his wife is light-shadowed, but explained that she is really only more easily frightened and that those creatures she alone sees or hears are only the product of her fears. There was some inconsistency in his reports, but inconsistency in regard to beliefs is one of the most outstanding characteristics of the rural Greek —and of most human beings, for that matter.

We are unable to resolve the question whether Dionysios believes himself to be a sorcerer. He does consider himself a magician, and a good one. His ability to bewitch and control the moon is greater, by his own admission, than that claimed by anyone else in the entire region.

He is certainly aware of black magic, and we know he has had an opportunity to learn it through his kinship and apprenticeship to the son of Miltiades. Others of stature in the community and region believe he is a sorcerer; they accuse him of having caused the death of the members of a whole family—no light matter. Certainly Dionysios occupies a deviant social role in the eyes of at least some of the villagers; the stress they place on his doing "women's work" imputes a role of social hermaphrodite to him, and the fear in which they hold him may well be related to his usurpation of the potency of womanhood. (There is no imputation of homosexuality in their talk.) In any event, Dionysios is a magician engaged in healing work, a specialist in illnesses of the moonstruck, a secretive fellow who is cordial enough socially, and a powerful, fearsome, awe-inspiring man-woman in the eyes of others.

VLACHOS THE GOD

The greatest komboiannitis or practikos in Greece is Vlachos ("the Shepherd") of Athens, called by some "Vlachos the god." We visited Vlachos on three occasions. Once we took a Saracatzani woman and her husband to him on their request when the woman had broken her ankle; a month later we returned with her for the removal of the plaster cast; and shortly afterward we went to observe him again at his work.

The compound in which Vlachos has his tiny office is on the edge of Athens. On each of our visits, 50 to 75 shepherds, peasants, and Athenian townsfolk were sitting about the compound. Many, of course, were relatives of patients who had been brought in; a few had sprains or fractures; others complained of arthritis; a few, it is presumed, came with ailments that were not orthopedic—for example, a Greek-American had come from one of the Western states to see if Vlachos could restore his sight.

The people within the compound immediately formed a friendly, talkative group around any new patient who arrived, whether by bus, donkey, or shiny new automobile. Those already waiting helped to make the patient comfortable near the door to the treatment room, and would show avid interest, asking the newcomer how the accident had occurred and showing a sympathetic concern with his story. Someone in the patient's family would knock on an unmarked door, under directions from others there, to be greeted brusquely by a man inside who took the patient's name and directed the inquirer to wait until the name was called. That wait was ordinarily from three to eight hours; not even a fracture case could expect to be seen out of turn. Sometimes, it was said, patients would arrive at 3 A.M. in order to be first in line. Vlachos himself may work from 8 A.M. until midnight, with only a one-hour break for dinner.

Most patients appeared to accept the long wait and the attendant pain in a spirit of stoical unconcern. When they come to Vlachos they know what to expect, and as people who have had pain before, they are unlikely to be anxious about it as long as the reasons for it are known. The writer Lawrence Durrell observes in *Prospero's Cell* that the Greeks are very brave when faced with physical pain or suffering if the cause is known to be mechanical, as in a fracture, but they may become intensely anxious, hypochondriacal, and disorganized in the face of a minor discomfort of unknown origin.

The patient probably derives considerable psychological support from the waiting group itself. Their immediate demonstrations of interest, their sympathy, their easy, natural manner with one another, and the almost protective cohesiveness of the group all provide the sociable Greek with a situation in which he suffers with others rather than alone; his distress is a community affair. The groups in Vlachos's compound, on the three visits that we made, were impressive in their warm cordiality and active acceptance of all newcomers. The distrust that one so often sees among Greeks meeting one another as strangers did not exist.

Should someone become dissatisfied with the long wait, he is given no special attention. During one of our visits, one patient's sister grew tired of waiting; her angry protests were met with a gruff rebuff from the physician through whom Vlachos's patients are channeled. She would just have to wait her turn, he told her.

A physician acts as receptionist for Vlachos in his outer office because, so goes the tale, of the charges brought against Vlachos by the health authorities for practicing medicine without a license. These charges led to a trial—itself a subject of dramatic stories—and the stipulation that his practice must be under the supervision of an accredited physician. This stipulation has also led to the installation of an x-ray machine, which is to be used in diagnosis and evaluation of treatment results. The supervising physician reports that in an orthopedic case the decision may be made to treat the patient medically rather than through hand practika. In such a case, the patient is taken to the physician's own clinic, located elsewhere in Athens. In the meantime there exists a supply of local and general anesthetics and other drugs in the outer office, available for use by both Vlachos and the physician. We did not see any cases in which they were used.*

Working in the same outer office in which the physician first receives the patients and jots their names on the waiting list is a gigantic man, the son of Vlachos. He also knows the methods of hand practikos, having learned them from his father, but he specializes in massage. The scene in the anteroom, on one visit, was as follows: In a room opening on the compound, no larger than ten feet square, stood a massage table, a small desk covered with boxes of anesthetics and drugs, and five or six simple wooden chairs. Seated on the chairs were four women, one an older person with a greatly swollen foot. This woman and her daughter, a husky, friendly, well-dressed Athenian girl of about 35, were next in line for Vlachos. The other two women were also waiting, but their complaint could not be deduced from their appearance. Also in the tiny room was a peasant man, standing, and his little boy. The supervising doctor, dressed in an open collar, fine silk shirt, and casual slacks, conversed with us (two team members). On the massage table lay a woman, very heavy, about 70 years of age. She was nearly naked, for her buttocks, covered with torn cotton panties, and her ancient brassiere were exposed under a small sheet. She was moaning and groaning as though in excruciating pain as the masseur, the son of Vlachos, massaged her. The doctor, in response to our inquiry, said the old lady had "spongeoarthritis" (ankylosing spondylitis?). No one in the room paid the least attention to the old woman's loud groans or nearly naked state. Apparently, the standing man and little boy were with the old woman, but neither of them spoke to her while she was being massaged.

Vlachos himself is a man of 80 who dresses in an old shepherd costume, which he wore when he watched sheep years ago before he became a full-time healer. The costume consists of fustanella, pom-pom shoes, hat, and embroidered jacket. He is straight as a ramrod, his face utterly impassive and unperturbable with a look that is both stern and gentle. He has an air of complete strength and confidence, an air that implies the wisdom of the old, a closeness to nature, and the power of a patriarch. His patients respond to this image; they are quiet and respectful, awe-

* It should be noted that most drugs for which prescriptions are required in the United States may be purchased without prescription in Greek pharmacies.

struck and obedient, and they address him as "my father." He is, by any measure, a forceful personality.

His treatment room is tiny, no more than eight feet square, with two heavily curtained windows, one opening onto the compound. The walls are covered with dozens of silvered ikons sent to him by former patients —the same kind of ikons that one finds in churches as gifts rendered to the saints in return for cures. There are also several hundred silver ex-votos of arms, legs, hands, feet, and torsos, which are hung on a string along the wall above the old bed that serves as both couch and treatment table. These too are gifts to Vlachos, promised by patients and given him when the desired cure took place. The office, then, is a shrine; except for Tenos, it probably has more ex-votos and silvered ikons than any other holy place. Vlachos he is, and a god indeed.

Vlachos sits by the head of the couch, erect on a wooden chair. To his left stands the couch; to his right in a corner is a small table laden with bandages, plaster, scissors, and other tools. At the other wall, to the right of the corner table, sits a small woman, his cashier, who keeps the cash box in her lap. There is room in the office for little else; the two chairs brought in for us to use while we observed his work completely filled the remaining space.

We will describe two treatment scenes. The two of us were introduced to Vlachos as a young man left, carrying with him a four-day-old child. The father explained that the infant's arm "had been broken by the doctor who, at delivery, had pulled the infant out feet first." (Kostas the Turk in Doxario had reported that fractures during delivery were very frequent.) The father showed us the x-ray photograph reputed to demonstrate the break.

We were invited into the treatment room at the same time the husky Athenian girl assisted her mother to the couch. Neither woman made any objection to our presence, nor were they given the opportunity to do so. The daughter was respectful to Vlachos and smiling to us; her mother was in too much pain to be attentive to anyone.

As soon as the old woman was seated on the couch, immediately to the left of Vlachos with her daughter on her own right on the couch, Vlachos grasped her ankle and began to probe with his large muscular fingers. The woman began to scream and thrash about, her skirt flying high and her arms flailing. Her daughter addressed her sternly and then, seeing no quieting, put a wrestler's headlock on the old woman. Soon another woman came in from the outer office—one of the waiting patient's relatives, it appeared—and held the thrashing leg. Vlachos maintained his firm grip on the other leg and continued to probe without comment or notice. The woman continued to scream horribly, crying, "Oh, my father! Oh, my father!" Vlachos spoke softly without any change in his impassive expression, "Do you want me to treat your ankle?" "Yes," she replied. "I will be as gentle as I can," he said, and continued to probe. The woman resumed her screams. The daughter remained completely unperturbed, saying only a few words to her writhing mother and taking time out to discuss with us how she herself had come to Vlachos some years before when she had hurt her own ankle. She showed no distress or disapproval about her mother's cries, nor did she ask Vlachos or the doctor to give any anesthetic.*

As soon as Vlachos had completed his examination, he stopped and turned to his small gray-clad cashier. He waited while the cashier told the daughter to pay 432 drachmas (about $14). The daughter reached into her purse and paid out the money, which was put into the cash box. When it was inside, Vlachos resumed his work, applying plaster and bandage to the ankle. When the work was done, we asked the daughter if an x-ray had been taken. She said it had not. The supervising physician then came in, and the daughter asked him about getting an x-ray. The doctor replied she could get one taken if she wished. At this the daughter picked her mother up in her arms and carried the now quiet and pleased woman several hundred feet across the compound to the x-ray room.

During the discussion of the x-ray, the daughter had mentioned to us that the supervising physician had told them, when they were seen in the outer office, that the old woman had a fracture. "But," she went on, "Vlachos says it is only a sprain." The physician, who was listening to the discussion, commented, "So it's a sprain then; it's whatever the old man says it is."

On another occasion, a Greek-American who had returned to Greece

* Zbrowski's work on responses to pain in various cultural groups shows how persons from the Mediterranean cultures (Italians, Jews) are expressive and noisy in response to pain, while Anglo-Saxon peoples try to remain quiet (75). We suspect considerable intra-group differences, for the Saracatzani woman we took to Vlachos with the broken ankle was absolutely quiet, only gritting her teeth, while many of the patients of Vlachos whom we saw—or heard—were fully expressive; their shouts could be heard hundreds of yards away.

for a visit to Vlachos was being massaged in the anteroom. His nearly naked state was of no concern to him or to the other patients and relatives who talked and waited in the room. He had had neurosurgery for a herniated disc in the United States but found he still had pain; after a massage by the son of Vlachos, he reported he was free of discomfort. During his conversation a male patient in Vlachos's treatment room was shouting with pain. No one in the anteroom or courtyard took any notice; all continued to chat amiably, occasionally repeating their words when a particularly loud shriek had drowned out the conversation.

When the patient, a husky man of about 40 with his broken arm set, left the treatment room, now smiling and talking easily, a young woman and her husband entered. The woman turned pale, tensed, clenched her fists, and moaned slightly, but she was intent on not crying out. Vlachos diagnosed a broken great toe and began to set it. At this moment the supervising doctor brought in x-rays that had been taken of the foot and, without comment, gave them to Vlachos. Vlachos held the x-ray transparencies to the light and with evident interest studied them. Apparently they confirmed his clinical diagnosis, for without comment he returned to complete setting the bone, mixing the plaster, and bandaging the foot.

The supervising physician was happy to discuss the reasons he believed were responsible for Vlachos's great reputation and success. Vlachos, he said, uses a special plaster, its formula a great secret, which results in quicker healing of the fractured or sprained joint. In addition, he charges less than doctors do; moreover, the people have complete faith in him. His fame, the doctor said, is worldwide. As an example he cited the recent visit of a homeopathic physician from Los Angeles who had studied with Vlachos for 20 days. To the surprise of Vlachos, the homeopath had used a magnetic ball to locate fractures, running the ball up and down the limb until a magnetic response signified the fracture point. Although the supervising physician made no direct comment about this method, one inferred that he and Vlachos considered it a form of quackery.

When asked to compare Vlachos with a doctor, one of the Saracatzani said that Vlachos was not an educated man, that he was not polite and did not say "please" like a doctor might, but that he was frank and "spoke like a sword," telling one in a few words what was wrong and what must be done. He did not explain things in detail. He had said, for example, to the Saracatzani woman we had taken to him, "Your ankle is broken; I will heal it. You must rest it for 30 days. Then come back to me." That was all he said. The Saracatzani woman had followed his instructions to the letter. The Saracatzani, it would appear, are more comfortable with one who is a shepherd like themselves, whose knowledge, like theirs, is built on the bodily mechanics of the sheep with whom they are familiar. They enjoy going to a man who is a folk hero; witness the stories they tell of his triumphs over the medical doctors who brought him to court. In addition, just as Kostas had said, there is no language barrier between them, no need for formality, no chance for the shepherd to have a feeling of shame or inferiority, as he does when he compares himself with the educated and sophisticated gentleman who is the usual physician.

Thrift should by no means be overlooked as a motive for peasant behavior. Some stress Vlachos's low fee as their reason for loyalty, as in the following account:

> Vlachos is a remarkable healer. With him it is not as with the doctors, because you don't have to go more than once or twice at most. And when you go back for him to take the plaster off, he does not get paid; the only thing is that the assistant asks you to leave whatever you want for "the grandfather." In the old days you left whatever you could afford, but now that the government has made him keep an x-ray specialist, he gets 300 drachmas for the [first] visit. The doctors forced this because they didn't like his competition. Since his name is Athanasios, many people send him ikons of that saint. When my wife broke her leg and my son his ribs, they both went to him and were cured. They would have needed a great fortune and too much time if a doctor had taken care of them.

Another source of success for Vlachos probably is found in his effort to include the family in the treatment program whenever he can. The mother may be assigned the job of massaging the injured patient; the sisters may be given minor nursing roles. He utilizes and directs the strong wishes of the family members to participate in the care and healing of the patient, expanding on their own nurturing roles and family obligations. Unlike the physicians, who fail to make the family members partners in treatment, Vlachos and the other folkhealers implement the cultural roles assigned to the female relatives, making them feel more valued. Perhaps this helps to account for the case Dionysios described in which a mother asked him to heal her child even though the boy was already receiving medical care. Family participation is indicated in the following account by a Dhadhi resident:

> We respect Vlachos very much, for he knows things that the doctors don't know. When our son broke his leg we took him there, and Vlachos tied it with boards and told him not to move it. When Vlachos finally took the boards away, our boy couldn't walk, for the leg was stiff. We took him back again and Vlachos massaged and jerked the leg; the pain was very great. But his mother was to continue the massage at home. She did this, and then she helped him walk around the room. When he was better, she helped him walk around the neighborhood until the nerves were freed.

In reviewing Vlachos's reputation one cannot discount the triumph, which he demonstrates, of the traditions of the rural folk over the city ways. By employing a subordinate physician, he has, if anything, gained stature. The inferior-feeling, downtrodden, envious peasants must get some vicarious pleasure from knowing that one of them bosses an educated city man about; the shepherds enjoy the story of a shepherd's winning a battle against the government authorities in court; the peasant chuckles over the tale, so often told, of Vlachos's cursing, commanding, and curing the King of Greece.

We conclude that there are a variety of reasons for the paramount position of Vlachos vis-à-vis medical authorities. He represents the shared, the familiar, and the understood. His social position poses no threat to philotimo and requires no awkward façades. In addition, he is probably a skilled worker and a conscientious one who inspires confidence, one whose words are clear and easily understood.

13

The Priests

Within the region of the three communities we studied there are three priests. All of them are married. Two of them live in the town of Doxario, the third in the nearby village of Spathi. They are very dissimilar men. One, Father Dimitri, is a handsome, white-bearded man, exuberant and powerfully built, who was a butcher's apprentice in Doxario until the villagers held an election and selected him as their priest. His uncle had been a priest before him; when he was a little child, the uncle had held him on his knee and called him Father Dimitri. This family expectation may have helped direct him to the priesthood. After his election he was required to go to seminary for a year, and since his return he has resided in Doxario. He is known as a jolly fellow who loves to drink, dance, and shoot.

The second priest, Father Manolios, is a small man, energetic, thoughtful, and spiritual. He is from Naxos and feels his isolation from the people of Doxario, who consider him a stranger. He, in turn, frowns upon their irreligiousness, comparing them unfavorably with the more religious people in the outlying hamlet of Dhadhi who, as refugees, are strangers in the land as he is. Father Manolios is an ascetic and intelligent man whose devotion shines from his eyes as he speaks.

The third priest, Father Vassilis, is fat and sleek and not above ostentation. Many villagers, especially in Dhadhi, seem to fear and hate him; a few say he is the innocent victim of that ubiquitous envy, the poisonous peasant desire to destroy the life and reputation of anyone who succeeds in rising above the rest. This priest is called by many names, most in themselves being charges of heinous crime. When they do not speak of him in those terms, they simply call him Father Terror. As observers, we were not in a position to evaluate his conduct, past or present. What he was accused of was so frightful, so inhuman, that we think it best to avoid repetition of the hearsay; we limit ourselves to the observation that villagers whom we considered trustworthy swore to his crimes, while others, equally trustworthy, defended him against the scandalous rumors, saying his character was unimpeachable. He lives in Spathi.

Despite their individual differences, the priests have much in common with one another. Each wears the traditional garb of the Orthodox Church, which consists of a long black gown, a high cylindrical hat, and a flowing beard and long hair. Each is a magnetic, dramatic person, and aware that while he walks the streets of Doxario the eyes of the villagers are upon him. These men have their faith in common, and proudly carry the great religious and political traditions of the Church of Byzantium.

The priest in rural Greece has several functions. In addition to maintaining the moral order and religious interests of the community, he helps to keep the villagers unified. As one who is divinely consecrated and theologically trained, he has great magical power; he has the capacity to direct the imposing strength of God and the saints for the welfare of the Orthodox community, to protect its members from danger, and to invoke the divine power for the good. It is in this role as magician that the priest leads the fight against the antagonists of the divine powers: the devils and their human allies.

One function of the priest is to heal, and the priests consider their healing capacities to be of very great importance. Their consecration, their role as intermediary between the villagers and the divine beings, and their knowledge of the means to manipulate the divine powers are all directed toward the healing of afflictions. Their healing power is described as being of two kinds. One is direct intervention whereby they use their rituals and prayers to direct power to the good. They exorcise devils that may have possessed an epileptic or insane person, speak incantations to ward off dangers threatened by the evil eye, and counteract sorcery and black magic with a more potent and more effective good magic. Good magic is more potent because it is endowed or consecrated by God, the saints, or the Panaghia, and more effective because the person using it, the priest, has himself been consecrated and can control the magic.

The other kind of priestly healing power is indirect intervention. It relies on the power of faith, which is said to reside within some individuals, although not within all, since some apparently are born with a greater capacity for it than others. The internal faith must be directed against illness by the believer himself. His ability to pit his power against that which has bewitched him—inadvertently, as in the case of the evil eye, or malevolently, as in the case of sorcery—can be helped by the priest, who with his own faith helps to concentrate the believer's power and may facilitate its growth. The notion of indirect intervention may also involve the merging of the power of the priest with that of the faithful, bringing them together to fight affliction.*

The devices used to control power vary. At the simplest level they include expressions of intent, the direction of wish and power expressed through concentration or elaborated in words or prayer. These may be formalized in the reading of written prayers or exorcisms which can be found in books; for example, a specific blessing (efchi) is read over a person who has a headache; another is read for a stomachache, and so on. Expressions of intent may be augmented by rituals that bring additional power into play. These rituals include burning incense, making the sign of the cross with consecrated olive oil, holding ikons, sprinkling holy water, invoking formularies in liturgy, which may bring several priests together in joint endeavor, and holding special services in the house of the ill or in a church. Further power can be sought by appealing to God, the saints, or the Panaghia to intervene directly. Such intervention is presumably invited in any priestly endeavor, but the strength of the intervention of deities is related to the frequency of the use of their names during the ritual, the number of persons petitioning, and the intensity with which they are petitioning.

FATHER DIMITRI

Father Dimitri was very happy to discuss his healing work. He contended that the families of the ill often call for the priest before calling in a doctor. They do this because "they have more faith in the effectiveness of the priest." He engages in healing willingly, for he finds that with faith in God healing power comes easily, and that afterwards one feels elevated and satisfied. He grants that there are dangers involved, "for

* These same psychodynamics occur (see Chapter 10) in the physician's belief in the need for the patient to have "faith."

there is magic present," but "when I represent the church, no danger can reach me."

Father Dimitri is aware of the limitations of his healing powers, saying that in some serious illnesses, or in illnesses requiring an operation, the priestly healing will not help. On the other hand, there have been some gratifying cures:

Some years ago, an unbaptized baby girl was hopelessly ill, according to the doctor's diagnosis. I was called to baptize her, and the doctor asked me to do it the Catholic way [sprinkling] because dropping the child into the water the Orthodox way would be risky. I did not consent to it since I believe, as others do, in the healing powers of baptism. I did it in the Orthodox way. The child became well, and today she is a healthy young woman.

He told of another case:

I had a friend whose child's eyes hurt. The child was taken to the hospital and was about to be cured [probably of trachoma]. At the same time I asked my friend to have a service in the Church of Aghia Paraskevi [St. Friday, who heals the eyes]. Before having the service he bought a pair of silver eyes [ex-votos] as an offering to the ikon. On the way to the church he lost both the offering and his glasses. I assured him that he was going to find them anyway because the saint would help. On the way to his house from the church [after the service], we saw the offering gleaming on the road. Many cars and people had passed it, but it had been left untouched. The child's eyes became well after the service.

Father Dimitri, like other rural people, does not put all his healing eggs in one basket. In addition to church services and special healing rituals, as for example the "efchelaio" (special prayer or blessing), which may be said in the house of someone seriously ill and which uses consecrated olive oil that has special powers, he recommends the use of physicians, because of their demonstrable knowledge and skill, and the use of wise women or other folkhealers as well. On one trip into the mountains he gathered some healing herbs for us; on another occasion he told us how his wife's rheumatism had been cured by a wise woman who knew about herbs. She had made a mixture out of the fat from the kidneys of a pig, olive oil, camomile, white wax, mastic from the island of Chios, and the herb rue. This mixture, applied and left to dry while his wife sat in the sun, relieved her discomfort. However, she is not well, so each day she drinks a bitter tea of mountain herbs, said to be good for her kidneys.

FATHER MANOLIOS

Father Manolios was the best informed of the priests and the shrewdest observer of village life. We spent many hours in his company, seated over coffee in his house or at ours, and traveling with him to holy places in the region and to remote mountain houses, where he performed special services to protect the householders from the dangerous supernaturals who inhabit the lonely wilderness.

He recalled that he had dreamed of becoming a priest from his earliest childhood. After escaping from a carpenter's shop in Naxos where he had been apprenticed, he entered a monastery at the age of 12; there he put himself under the protection of the monks and began to practice the ascetic life. He was, he said, urged on "by a power," and while in church he feels the potency of the ceremonies upon himself. His healing powers are derived only from God; they can be effective, he feels, only insofar as he can transmit his faith to the people themselves.

He recognized a variety of illness causes, mentioning the climate, the hard life of the villagers, the disturbed emotions attendant to the violence, lust, and lack of love in family life, dietary and malnutrition factors, flies, lack of cleanliness and public sanitation, and the presence of swamps. He was the only healer, besides the physicians, to assign alcohol a role in illness, saying that alcoholism is a local problem and leads to mental deterioration.

It is his belief that those most vulnerable to illness are the refugees who came from Asia Minor and who have not adjusted to the humidity of the local climate (in winter); the children and the old people, who may get less to eat or have less adequate care; children who are reared by mothers ignorant of hygienic standards; postpartum mothers who, during the 40 days after delivery, fail to be hygienic; babies who, because they are not clean, are exposed to flies, and babies who are not nursed; girls who are about to be or have just been married, because they have been cursed by their mothers-in-law; blasphemous people, of whom there are too many in the Doxario region and who must expect a bad end; a few who are the victims of sorcery aimed at them by women whose animosity and unfulfilled desires for marriage lead them to engage in black magic—for example, men who are cursed by girls whom they promised to marry but did not; those who marry their cousins and contribute to degeneration; and perhaps some who believe they are victims of the evil eye.

Father Manolios is especially concerned over the many disorders that he attributes to emotional distress in the lives of the patients. He said these people become upset and then ill for good reason. He pointed to the poverty of the people in Panorio, saying it is understandable that they should worry a great deal over economic matters. The people in Dhadhi are not poor, but they suffer from their own financial mismanagement. They do not know how to control their expenditures or plan their affairs. Furthermore, many of the men spend all of their income on alcohol and gambling; other families have too many children and exhaust their resources in giving dowries, after which the children move to Athens and may not support the needy parents.

Family relations are, he feels, not good in Doxario. The husbands are indifferent to the wives and beat them; the wives do not know how to rear the children; the husbands are not at home because they are out getting drunk, and consequently family discipline and cohesiveness are disrupted. There is frequent incompatibility between the husbands and the wives, with parents often setting bad examples for their children by using poor language or by the father's beating the mother in front of them. The children in turn grow up to be disobedient, associate with bad companions, and, as adults, continue the cycle; they lead lives in which there is no link between the church, the school, and the family, and in which religious and moral values are regarded indifferently. This state of affairs is less commonly found in Dhadhi, where the villagers are more devout. Whatever antagonism the villagers there feel toward the priests may be justified. Father Manolios said that the Dhadhi villagers have had unhappy experiences with money-minded priests, ones who are not interested, he says, in the salvation of souls.

Father Manolios feels that there is a gulf between him and the people of Doxario. He wants to elevate them, but the ignorance, indifference, and difficulties are so great that he is discouraged. One would hope the authorities might help, but no matter what the church and the local government attempt, the task remains difficult. Sometimes the authorities themselves are not interested, and corruption occurs among them to discourage one further. The people are quarrelsome and litigious, forever taking one another to court over the use of fields or fence and property lines; they are overly occupied with envy and the evil eye. Father Manolios is concerned that there are such differences between himself and the people:

Take the matter of epilepsy. The people here believe that epileptics are paying for the sins of their fathers or families. They think that the intervention of the church will drive the demons out, for demons are also believed to cause it. Besides sin, people attribute epilepsy to the Bad Hour, especially when the person has done some wrong and the devil has found him weak at that moment and taken hold of him. So, they think it is a demon, but I know it is a disease of the nervous system.

In regard to his healing work, he said, "I can only help people who have maintained their reason. Those who have lost their reason I cannot aid. For example, schizophrenics and epileptics cannot be helped."

Healing work, when the prayers are effective, is exalting, but it can only be exalting if the priest takes no pay. A priest who would take money to heal is unworthy of the calling. There is a strong effort invested in healing, said Father Manolios; one feels a certain "spiritual lassitude" afterwards. Asked to explain, he said:

This spiritual hangover that overcomes me is due to the transference of power to the relatives of the sick person so that they will believe in the power of God. [Then] I am able to present myself in front of Him as His true representative. I am not exposed to any other hazard or danger, since the sign of the cross and the divine element in my work protect me from any kind of danger. My health remains in an excellent state and is not affected by the lassitude that comes from my healing efforts.

During the ordination ceremony, the Bishop puts in the hands of the prospective priest the bread that symbolizes Christ's body. By doing this he puts a burden on the priest's shoulders because he considers him qualified to transform the bread into flesh. Whatever the priest wishes at this moment, when he has pure faith in the power of that moment, will happen. This was the case with me [leading to my power to heal others].

He told of a case he had cured:

A teacher came to me at midnight because his two-year-old child was seriously ill with high fever. I went there immediately. You must know that there will be no cure if the priest complains about the hour; he must be prepared at any time he is asked. It is better though for him to fast [to purify himself] if he knows in advance that he is to be called. The child was unbaptized, so I made the sign of the cross in the air and said the other prayers. By the following morning the child was playing in the garden.

There is a case in Panorio in which he does not expect a cure:

There is a matchmaker who wanted a girl, the daughter of a woman I was called to treat, to marry someone from Spathi. Her parents wanted to give her to someone who would be of a higher social position [the parents are shepherds]. The matchmaker lost the business because of their aspirations [which resulted in a match with a boy from a distant town]. The matchmaker began to talk to the mother, stirring her up, telling her, "Ah, now you are going to lose your daughter. She will go far away and you will never see her again." The mother became worried and slothful; she turned inward upon herself, came to believe everyone was against her; she refused to talk and was irritable. She had the constant thought that her daughter would be lost to her. Her husband called me to reason with her. I think I may have influenced her, but I am not sure how much. Nevertheless, I told him I could help his wife, and told him to call me at any time. I read a special blessing over her, but she is a hypochondriac. I cannot help her if she does not maintain her reason.

Finally, he told the story of a miracle he witnessed:

Early one morning, I was in my home town of Naxos when a car stopped outside the church and a lady from the town, an acquaintance of mine who was also an acquaintance of the town pharmacist, came to me and told me a dream she had had. The night before, a lady had come in her dream and asked her to come to me in church and to take from my church an ikon representing the Annunciation. She was told, in the dream, to put the ikon on the pharmacist's child. This child, by the way, was very weak, he looked undernourished, but that was not his trouble; he was merely short and weak. I looked around and told her I did not have a small ikon of the kind she indicated. The only one I had in the church was very large. My wife was present, and asked me if the lady might not mean the small ikon of the Annunciation that we kept in our home, one that was a gift to us from the holy island of Tenos. I did not wish to give her that ikon, so I sent her away.

I went home. I kept the ikon on a night table by my bed. I noticed that its doors were closed—it was a wooden three-dimensional scene, with little doors that opened, enabling one to see the Annunciation portrayed within [a triptych]. I tried to open those doors, but no matter how hard I tried, they would not open. I tell you I shake when I remember it; the hair on my arms stands on end just to think of it. Well, besides the doors' being closed, I saw that the ikon was perspiring; there was water coming off it. I assure you, the following day I took that ikon to the child of the pharmacist myself; I put it on his body, and at the same moment the ikon doors opened and the perspiration ceased. And from that day the child developed normally. This I saw myself.

In these accounts we can discern the elements of purification, power transfer, magical intention transformed into physical effect, the apotropaic value of religious signs, and the promised protection by the deities of the health of the priest. We also observe the folk element of nonpayment for healing work, the significance of dreams, and the intervention in matters of health and illness by the saints. One of the stories illustrates the priest's belief that faith healing can occur only in rational persons; although he may reassure the husband of his ability to help the wife, the priest himself is not at all convinced of it.

FATHER VASSILIS

Although we were with Father Vassilis on many occasions, it was difficult to lead him to discuss his own life or activities. Suavely, he managed to keep discussions in the abstract, and to turn away any reference to his own career or observations. Indeed, the only time he spoke of contemporary events was when he became incensed over the educated men in the priesthood who, by virtue of their theology, he said, had forgotten Christ's injunction to "follow my word." It was his opinion, as well as Father Dimitri's, that the educated priests had lost the faith; they both expressed disappointment in the movement of the church toward greater theological training for the priesthood. It seemed to be their feeling that faith—in this case, the power to influence natural events through belief in the capacity of the saints and God to effect such influences—accrued only to the simple folk. As people became educated, their knowledge became a barrier between them and their power to believe in and influence the divine figures. It was implied that the healing powers were more often found in the simple priests, i.e., men of peasant stock who had not abandoned the rural traditions for city learning.

Father Vassilis said that within the preceding three months he had been to Dhadhi on only three occasions to heal the sick; the reason, he said, was that Dhadhi was not really his responsibility, and that the refugees there were supposed to call in either Father Dimitri or Father Manolios. The villagers do not get along well with either of them, and that is why they have departed from the older custom of calling for the priests. (In contrast, Father Manolios had said that the Dhadhi villagers did not get along well with Fathers Dimitri and Vassilis, but that they got on well with him.)

Father Vassilis said he does have responsibility for Panorio, and that he has gone there four or five times in the last few months in order to heal the sick. He will say prayers, bless the house and the people, sprinkle holy water, make the sign of the cross with oil, burn incense, and so forth.

It was his opinion that the power of faith, drawing on the protection of the divinities, is greater than the power of the microbes that cause disease. Accordingly, he does not hesitate to take communion wine from the same spoon that the sick villagers use. He is exposed to contagion in this way but has never become ill as a result. He said that the ritual of communion provides total protection to all who participate in it. For that reason he feels no hesitancy in putting the spoon with the wine in the mouth of an infant, even though he knows that the person who has just had the spoon in his mouth has tuberculosis. "Although infants are especially vulnerable during those first 40 days," he said, "during communion they are protected from disease by holy faith."

He told us the following stories to demonstrate the healing powers:

It happened in the Church of Aghios Dimitrios [St. Dimitri] in Tenedos, where I was for many years. One day a group came to my church to ask me to have a special service. While I was reading the gospel during that service, I heard the sound of teeth chattering and saw a young man falling down before me. I did not interrupt my reading, but as soon as I was finished I went to the man and knelt. I said a prayer, and crossed him four times. Immediately, the young man stood up. When I asked his family what was wrong, they told me he suffered from epilepsy, and that usually each crisis [seizure] lasted 30 minutes. This was the first time that it lasted such a short time. From that time forward, the family visited the church each year on the day this miracle happened to their son. Always afterward the duration of his crisis was shortened. Now he is able to work. I should add that Aghios Dimitrios is a church with a reputation as a place where people are cured.

February 10 is the day the Church of Aghios Charalambos in Euboea celebrates the memory of the saint. The church has the reputation for healing since St. Charalambos has healing qualities. Many sick people visit it. Among them one day a deaf and dumb child was brought by his mother. He knelt along with his mother before the ikon. She was praying that the child would speak. All of a sudden a cry was heard: the child cried out, "A white dove." He had seen a dove that came to him, a sign, and he cried out, no longer dumb.

It was in Tenos as the holy ikon of the Panaghia was being carried over the sick people. Suddenly there was a cry—where it came from no one can say—"Oh miracle," and at that moment some among the sick who had been dumb and paralyzed were completely cured.

The cures that Father Vassilis cites are each miracles, occurring in holy places as the saint intervenes. The priest need not play any active role; one's presence in the holy place may be sufficient to bring about a cure. In each case the church is one with a reputation for miracles, and the supplicants who come pray with this knowledge. The cures described by Father Vassilis remind one of those reported at Lourdes. Insofar as Father Vassilis's accounts can be verified by impartial observers, it may be assumed that social and psychological factors, of the kind postulated by Jerome Frank (24) in his review of religious healing phenomena, are in operation.

14

The Administrative Setting

There are certain broad national factors in Greece that affect the health of the rural people. To begin with, Greece faces some great challenges associated with her present underdevelopment and her visible movement toward partnership in technology with the other nations of Europe. By Western standards, Greece is poor. The average per capita income is about $230 a year for rural people and about $450 for city dwellers.* Greece has few natural resources developed to yield income from their export; her people have not yet learned skills that would enable them to import and fabricate raw materials and export the finished products. The soil itself is too often harsh and unfruitful. The lack of high-income marketable crops means that peasants and shepherds cannot buy many manufactured products. Local industry, with small markets and high tariff protection, produces at high cost. The disparity between low income and high costs leads to low living standards.

The economic facts of life mean that there is not now enough money available to the country people to buy those goods or services they want or those we think are necessary to improve their nutrition and health. There are not sufficient funds to finance all of those ideal services that the government wants to provide: the necessary facilities and trained personnel for education, health care, transportation, research, and other endeavors to advance the public welfare.

A second challenge lies in the management of the limited economic

* As reported by Professor A. Papandreou in a personal communication based on his economic survey conducted for the Bank of Greece in 1962–63.

resources that do exist. How does one gather and disburse limited monies in a way that is most effective, protective, and humane? How does one plan fiscal policy and administer programs rationally to assure the maximum benefit for the land and its peoples? As a land rather recently embarked on self-government (the revolution brought freedom from the Turks in 1822, but some areas were not occupied by Greeks until 1912), Greece's administrative and political operations are not necessarily smooth and well organized.

Legal and moral controls to protect the welfare of the many against the self-interest of the few are still being evolved. For example, the low salaries of tax collectors and their culturally sanctioned loyalty to their own families rather than to a code of public conduct means that the government's tax income is considerably reduced as taxpayers bribe tax collectors. The bribing taxpayer is responding to cultural and economic forces: often his own business is marginal; he wants to provide well for his family; he has not yet learned any sense of national community or felt any obligation to contribute to the general welfare of abstract "others." His concern, even if he is a city dweller, is the same as the peasant's: one has obligations to one's family and immediate community; beyond those boundaries are strangers who must take care of themselves.

In addition, it has been suggested that the distribution of the tax burden may not be as rational or as proportional to income as it could be. As in any tax system, loopholes exist; often these loopholes favor those whose positions of wealth or power enable them to lobby most effectively with legislators who themselves are likely to be drawn from more favored economic groups—legislators who, because of their own family loyalties, may be open to unethical temptations. The consequence can be, for any nation, a tax system that is perceived as favoring the few and disproportionately burdening the many. However the case may be, it provides the "many" with a rationalization for their own grudging taxpaying and their self-interested withholding of taxes through concealment, falsification, or direct bribery.

Once tax money is collected, it must be spent wisely. This is a challenge for any country. In Greece there are several problems that arise in relation to allocation of tax monies. How does one distribute limited funds to the various sectors of the economy, each of which is needy yet interdependent with all other sectors? How does one earmark funds for the various branches of the government, harmonizing the needs of defense with those of development, the needs for public safety with planning for later income and security?

Politically, Greece has had a painful history of invasion, occupation, civil war, and civil strife. She shares borders with four nations (Albania, Yugoslavia, Bulgaria, and Turkey), each of which has been in conflict with her during the last century. She has only recently recovered from a disastrous civil war; elements of the rebellious party still exist within the country, and the elements for rebellion, considered in a socio-cultural sense, have not disappeared.* In addition, Greece is aligned with NATO, and must confront two hostile Communist nations on her border. During the civil war, the dissident mountain folk and dissatisfied urban elements received direct military support from these bordering countries. Because of the potential for international conflicts (and because of some unspoken but perhaps fond hopes for the recovery of her own "lost" territories), the Greek government states that it must commit a large portion of its budget, about 50 per cent, for security purposes.

Another element that contributes to the government's need to maintain heavy military and paramilitary expenditures is related to the factionalism and rebellious potential of the people, as cited previously. This rebelliousness, however, may be translated into broader psychological and political terms. Individually, the Greek male prides himself on his independence. Except in his relations with his own father and with dominant employers, he will vigorously exercise what he sees to be his political rights and defend himself against any threat to his philotimo if he perceives an abridgement of his capacities to do what he wants. Psychologically speaking, the rebelliousness displayed against authority figures is impressive. The often tyrannical father or employer is treated with the required respect, but the overt deference may only serve to conceal the insurrectionist rage boiling within.

For instance, the participants in political or community efforts often become engaged in violent disagreements that disrupt any possibility

* Dinko Tomasic (68) describes how the belligerent herdsmen of the Balkan mountains have maintained the civil-strife potential in these lands. Their patriarchal society, like that of the heroic age, following tribal traditions and adopting the Byzantine Ottoman patterns of "dynastic despotism, praetorianism, police rule, and venal bureaucracy," contributed to clan factionalism and warfare. By drawing on growing military forces or lining up with opposing outside powers with local interests, these groups have been militant in every Balkan land. Tomasic believes that internal harmony will come only when the standard of living rises and when military-police rule is dissipated.

of organization or progress. These disagreements may be accompanied by efforts to topple those who already have power or to prevent the ascendancy of those who might secure it. One sees here evidence of disdain for authority, of hazards to cooperative endeavor, and of a competitiveness and envy that prove so disruptive to organized action. Some have described the Greeks as anarchistic, but this is probably a side effect of their basic character. The demands of individuality and philotimo, the operation of envy and competition, the unsteady course between deferential obedience and violent rebellion, the lack of stable patterns for learning a means for cooperative endeavor, the heritage of concrete, family-oriented loyalties, and the absence of broader, abstract loyalty conceptions all lead to potential political instability.

Moreover, the law is ineffective in certain respects. Moral codes that guide conduct in the family and the village are traditional; they need not be written as laws because they are an integral part of the person. Neither he nor his fellows tolerates deviation from the honesty and decency that is inbred and that must be displayed in interaction with trusted persons. On the other hand, that law which is enacted by strangers in a parliament and which does not reflect the local moral custom is neither particularly salient nor mandatory. Father and the family command obedience; the priest, the community president, and the teacher may get respect; but the law and its enforcers are regarded as intellectually and emotionally foreign. The police recognize this; a police chief happily said, "Our law is elastic." A peasant said, "My obligations are right here in the village, not to somebody in Athens." The consequence of the artificial, external, unassimilated character of the law—of the policeman's being the enforcer of a foreign system—is that the wish to perpetuate the body politic under a rule of law subscribed to by all does not exist as a strong stabilizing social force in Greece.

When laws that are not part of the moral system are put into effect, increasing disdain for abstract "justice" may develop. The differential application of laws, the common corruption of relations between citizens and the police, the casualness of the courts in regard to enforcement, the conflict between law and moral custom, the mockery of jurisprudence as found in the hiring of witnesses, the overriding loyalty to friends or family instead of to principle, the extension of this loyalty in the operation of influence to corrupt the judicial process—all of these practical effects merely point up to the individual the distance between the real and the ideal, and sustain him in his own rejection of that ideal.

The consequences of the dominance of a local moral order instead of a more abstract rule of law, and the difficulties arising from psychological and cultural factors in the development of supra-family and supra-community organizations can lead to a sense of insecurity for any ruling group. As they face the local loyalties and the anarchistic tendencies of their people, any ruling faction might feel it necessary to protect itself through an elaborate police and military-security system.

To the extent that the ruling groups, whatever their labels, may not command widespread popular support—either because their own loyalty is to particular factions or because reliance on force generates further opposition—there will be increasing grounds for justifying the oppositional tendencies already in existence. Should there be added to these sources of public dissatisfaction any widespread belief that the government lacks interest in the welfare of needy population groups, then any envy-arousing, self-seeking, family-enhancing actions on the part of the leaders, or any publicized mismanagement or corruption in activities vital to significant population groups, will mean that the ruling elite may very well require force to maintain its power. This can be the case regardless of the excellence of its administration and even though it is supported by a majority of the population, for the minority will not be pledged to the idea that the majority has the right to rule.

Another set of factors exists to make any ruling group in an underdeveloped land that is undergoing rapid economic progress the object of anger born of frustration. We are speaking now of what Adlai Stevenson called the "revolution of rising expectations." When a people become aware of another way of life that they think is better, when they become aware that some members of their own society have already achieved that better life, so that they use these groups as a point of reference with which to compare their own status (43), and when they become aware that the society is moving toward a higher living standard and that they may legitimately allow themselves the hope of improving their lot—when these conditions exist, we can expect expressions of dissatisfaction over the slow pace with which improvements occur. This dissatisfaction takes the form of blaming the government, for the government does have power and responsibility, and in the mind of the peasant any failure to achieve desired ends must be due to negligence or design. The notion of technological advance as a historical process of immense complexity is less easily grasped and not very appealing emotionally. Like human beings elsewhere, the Greek peasant is likely to project his distress aggressively, finding a scapegoat (the pharmakos is a Greek tradition, after all) and blaming it for what has happened or failed to happen. It is simpler to think that the government has erred than to face the ambiguity of history and economics and find that no one is at fault. The weak man is fatalistic because he cannot control events, but he is not patient, nor does he believe that events cannot be controlled. He holds responsible whoever has power.

The Greeks have, in fact, made remarkable progress since the terrors of the German occupation and the civil war. But the rapid improvement has itself led to discontent. If changes do occur, why not all the changes one wants? If Athens has some fine new streets and grand new buildings, why not Dhadhi? Change encourages hope, hope that is often unrealistic, even grandiose. Lacking the knowledge that would help him to limit his expectations, and exposed to partial miracles that others seem to reap daily, the peasant asks, why not hope for everything? Here is an example:

One of the first people to be told that a public-health team was coming to Dhadhi, bringing doctors, nurses, and laboratory personnel, was Christos, a bright, ambitious, intense man of about 35. Our social worker explained to him that now anyone in the village who wanted to get a free medical examination could have it.

"Do we get free medicine too?" he asked aggressively.

"We will have some, but not a great deal," replied our social worker.

"What about x-rays?" he inquired sharply.

"No, we couldn't arrange that."

"How about free treatment and surgical care?" he demanded.

"No, we can't provide that, although if our doctors find something wrong, we'll do what we can to arrange further treatment in Athens."

He was very angry. "You mean we won't get free medical care for the rest of our lives?"

Our social worker paused and looked at him smilingly. "Don't you think you're carrying things a bit too far?"

Christos glared at her. Slowly his scowl changed to a broad grin. "Well, if you're bringing an English miracle, why not go all the way?"

Not everyone can laugh at his own expectations, as Christos did. Many Greeks become militantly anti-government; others, probably the majority of the young men and women who want many things but must settle for little, become deeply pessimistic. They overreact, despair, criticize every advance as insufficient, and let their pessimism run wild, destroying that newly generated idealism that their land needs so badly. Those who are pessimistic in the extreme are ready to retreat to the old patterns of corruption, self-interest, and apathy. Those who remain militant speak of the overthrow of the government they blame.

These are some of the complex processes—psychological, cultural, economic, and political—that threaten the stability of governments and lead the party in power to commit a large share of the national budget to securing its protection through police and military operations. Because a large share of the national budget goes to police and military operations, investment in development and public services suffers. For example, in 1962 the budget for government health operations was approximately two per cent of the national budget; that for public education—at all levels—was another two per cent.

The heavy commitment to protective, rather than productive, expenditures need not be lasting. Fortunately, social conditions are changing rapidly enough to enable one to hope for economic development, education, and enhanced communication leading to a larger sense of community responsibility, which will help to create a broader, sounder base for political power. Government policies also may be expected to change, reflecting enlarged resources and greater efforts dedicated to public welfare. It is fortunate, too, that Greece, as the mother of democracy, values highly the dignity of man, and that the individual Greek, for all his intense complexity, understands that men must live together constructively and peacefully.

Inefficiency in administering government agencies and programs is also responsible for the lack of funds available for health, education, and welfare services in Greece. Inefficiency occurs in large-scale operations in any nation; the problems of Greece are special only insofar as they reflect the special conditions for administration in a land that is partway between the traditional and the technical ways of life. The old patterns of family loyalty can lead to nepotism, to favoritism in hiring or contracting to help family, friends, or persons from one's own region, and to misuse of public funds as they are channeled to private purposes.

One also observes inefficiency arising from the insufficient supply of trained personnel. This is evident when the chief of a health bureau must do clerical work because no trained clerks are available; when ambulances break down because drivers are not trained to care for their vehicles, and when the drivers have collisions because they are not trained in safe or lawful driving; when faulty equipment is purchased because purchasing agents are unaware of job requirements or materiel standards; when laboratories stand idle because there are no laboratory technicians; or when untrained administrators fail to budget wisely and make capital expenditures on facilities that cannot then be serviced or staffed. These are but a few examples; the problem is widespread and affects every level of operation.

Related to the lack of trained personnel is the misuse of those persons who are well trained. Older administrators, suspicious of young people trained abroad, may prefer to hire an incompetent person rather than to "rock the boat" with vigorous young people who have new ideas. A more widespread misuse is a function of the present low salaries: many city Greeks work at two or three jobs in order to obtain funds to live, doing justice to no job and living a life of exhaustion. Government physicians working a four- or six-hour day six days a week for $100 a month, devote another four or six or eight hours to private practice or second jobs, leaving them no time for continuing professional education and no energy to invest in either work. Poor administrative practice may lead to misassignments, to a nurse's becoming a clerk or a physician's becoming, in essence, a laboratory technician.

The availability of trained personnel is inexorably linked to the adequacy of the public-school system and to the financial resources of the students' families. Many families cannot afford to send their children to secondary school or universities, not only because the costs of tuition and of maintaining the child while he is there are high, but also because as long as the child is in school he is unable to work and to contribute to the hard-pressed family. Naturally, there are strong pressures on the child to leave school and to contribute to the family income.

Public education in Greece is making rapid strides; the respect accorded to the professor and the teacher expresses the fundamental value the culture places on learning. The liabilities of the present school system are the same as those of other organizations faced with insufficient funds, untrained personnel, the discouragement of the idealistic young when confronted with the entrenchment—and corruption—of some of the elders in positions of power, and the discouragement of the elders who have worked so hard for so long and still see such a long way to go. The health professions depend heavily on the two universities for technicians, physicians, and basic scientists. The universities, in turn, depend on the secondary school system to ready the young for higher education. At both levels there are severe limitations: physical plants are inadequate, texts and curriculum are out of date, teachers are underpaid, and a sense of decay and discouragement too often prevails.

In consequence, the student culture lacks some of those attributes that would help to move the system forward or to fully exploit present resources. Students may come to be cynical about the learning process; they do not develop those habits of scholarship that assist in professional advancement. The lack of classroom space leads them to neglect going to class; inadequate libraries make them reluctant to read; insufficient funds prevent them from buying their own books; casual habits about punctuality and time mean that students *and* professors fail to appear for examinations; and insufficient laboratory facilities restrict the development of needed medical skills. The students develop a tendency to attend only those courses in which professors give examinations and in which a passing grade is required in order to obtain a degree; elective or secondary courses are let slide. In addition, many students must work to earn enough money to stay in school; in extreme cases, work outside school may amount to seventy or eighty hours a week. Standards of scholarship can be very low as a result.

The milieu of professors suffers as well. Fierce rivalry and envy impede scholarship. Only a few biological or medical teaching scientists have the time, the resources, or even the training to do research. Sometimes the professors' writings reflect outmoded concepts, ideas not supported by research, or even plagiarism. The fact that very few full professorships are available discourages the outstanding young men, who can never hope to achieve the recognition they deserve. Self-interest, favoritism, exploitation, bribery, and corruption also occur and may run rampant where inadequate controls exist. Respected schools may resort to bribing state examiners, especially if the examiners demand payment in order to certify even the competent students. Students and teachers, caught up in such situations, grow understandably discouraged and listless. It is a poor preparation for a profession in any health science, where one is likely to need confidence, trust, buoyancy, and optimistic initiative in order to render the best in scientific and humane service.

In Greece there are not many fortunate enough to have learned the science and art of management, and fewer still to teach it. Consequently, the very best-intentioned administrators lack the knowledge that might enable them to maximize the potentials of their working force. The organizational problems we have observed in health-service institutions include: (1) Detrimental effects arising from poor distribution of power—the failure to delegate authority and to make responsibility commensurate with it; (2) ineffectual communication—where coordinated planning did not occur and where operations in various divisions proceeded independently of, and even in conflict with, one another; (3) defensive orientation instead of task orientation—where individuals jockeyed to protect or enhance their positions rather than to get assigned tasks done; (4) inattention to employee morale—failure to understand the elemental techniques whereby employee job satisfaction can be increased and thus performance level increased; (5) failure of supervisory personnel to support their workers—occurring whenever an official fails to facilitate the work of subordinates, or to follow through on promises to assist them, or whenever he betrays trust, or leaves them to shift for themselves in work situations where his power to aid, pay, or encourage is needed; (6) employee unreliability, arising partly from the failure of the organization to define over-all work responsibilities and demand performance as a condition of employment—seen in absenteeism, favoritism, scheming and plotting against superiors, carelessness in work performance, petty larceny, etc.

In addition to these general problems, there are specific ones associated with hospitals and clinics where direct patient care is given. Since we did not do a study of these institutions, we can comment only on the results of random observations. Nevertheless, from these it would appear that patient-care problems in Greece do not differ from common problems elsewhere. As in other countries, failure to control the quality of care can be a serious matter. If no supervision of medical service through medical audit or tissue committees exists, if careful control at all levels of nursing service does not take place, patients are put in jeopardy. If records are not carefully kept, or if reportable diseases (smallpox, typhoid, malaria, brucellosis, etc.) are not reported by examining physicians, needless public-health dangers arise. These difficulties are, of course, associated with the kind of training, the facilities available, and the orientation of personnel to their patients and to one another.

Another cause of many of the personnel and organizational problems encountered in health agencies in Greece is the uncertainty with which most institutions operate. Only a few remarkable institutions, for example, the Bank of Greece, offer any job security. In most settings the employee lives from day to day, fearful that something may occur to threaten or deprive him of his livelihood. The future seems very uncertain; there is always the possibility of not being paid because of a paper-work error, of being made a scapegoat by an angry boss, of losing one's job on a moment's notice, or of being humiliated, rejected, accused, or disregarded. An employee is very likely to be anxious, superficially deferential, inwardly rebellious, and acutely defensive. The old peasant theme of distrust cannot help but emerge and be reinforced; the employee comes to feel that he cannot count on co-workers, subordinates, or supervisors to support him, keep their promises, or treat him fairly. Why then give them loyalty? Why put his shoulder to the wheel? Why not do what he can to protect himself and his loved ones against the multitude of distrusted others? Distrust and the lack of stable organizational operations cut deep into the effectiveness of individuals and institutions.

Perhaps one of the most striking characteristics of the people performing in organizations is their reluctance to admit responsibility for error. It is true, of course, that most human beings find it more comfortable psychologically to think well of themselves and to deny faults. The philotimo of the Greek male reflects this tendency. A man is proud, a man is master in his house, a man avenges insult, a man bows to no one, a man is *blameless*. Misfortunes occur and accidents happen—fate, the stars, a demon, the "bad hour," all may intervene to impede the desired course of events—but for a man to make a mistake, no! To run an organization in which every man sees himself a king, regardless of the realities of status and power, an organization in which every man seeks power but none will take the responsibility for the mishaps that occur—that is a management disaster. And disasters there are, daily. If a supervisor tries to assess a situation and lay the blame where it is due so as to avoid a repetition of the error, an uproar can be expected. Philotimo is a noble concept, making ordinary men grand and giving color and passion and strength to a nation; it is a warrior heritage. But in those complex institutions that evolve to meet the needs of technological societies, where individuals must be subordinated to create a series of interchangeable, predictable personnel units, individuality is intolerable, but philotimo is calamitous.

We have outlined a few general circumstances that have profound bearing on the provision of medical care to rural people. We shall illustrate these implications with a series of examples taken from our work or reported to us.

Example One: The chief of a public-health–education unit had worked hard to set up an information program in some mountain villages. He had aroused the interest of the local people, and had told them there would be a series of meetings where movies and demonstrations would be given. The people, ordinarily antagonistic toward government representatives, had abandoned some of their distrust. The unit chief returned to the capital to make final preparations. Just before he was to leave for the mountains with his staff and the movie equipment, he was told there were no cars available to transport them back to the mountain villages. Without adequate public transportation, they were unable to meet the villagers as they had promised. The villagers, it was reported, were much disappointed, and renewed their distrustful orientation toward government personnel.

In this example we see what happens when an insufficient number of vehicles is assigned to an agency; when a responsible unit chief has insufficient power to meet his responsibilities; when higher-echelon administrators fail to understand either the need for public-health education or the consequences to a program, to a region, and to morale that result from undermining the authorized work of a subordinate; and when communication is so poor that one unit makes plans that do not take account of the administrative disposition of needed equipment.

Example Two: An upper-echelon health worker approached a high government official, asking him to give consideration to an increased budgetary allotment to public-health work. The official declined, arguing that he could not see tangible benefits to be gained from public-health programs.

The limited scope of the government official's information and interest is combined with probably ineffective reporting of public-health achievements. Good training and broad interests are required for government planners; effective reporting and public-relations communications are in order for health and welfare agencies. It should be a matter of public knowledge that public-health efforts in Greece have nearly eradicated malaria, and that life expectancy has in 30 years gone up from 49 years to nearly 64 years for men and from 51 to 69 years for women. All responsible officials should know to what extent agricultural and industrial productivity are affected by worker time lost through sickness and by work poorly and slowly done because of debilitating disease. In our study of the village of Dhadhi, for example, nearly one-fourth of the working-age population had been chronically ill for over a year; half had been bedridden with illness during the last year; in the community as a whole, each man, woman, and child averaged 18 days—as a minimal estimate—in bed during the last year.

Example Three: The public-health doctor-nurse-scribe team arrived in one of the study villages two hours late and in a high state of agitation. One of the high officials had taken the car assigned to them, and they had finally secured transportation in an ambulance. The ambulance was to return immediately, for it had been assigned a run high into the mountains to get a patient seriously ill with a contagious disease. The team had been told the ambulance would drop them off, but they would all have to get back to Athens as best they could; no provision would be made for transportation home from the field work to which they had been assigned. They were very upset, one physician threatening to go back with the ambulance. He clearly expected us to betray him just as his organization had done. To the team's considerable relief they were not stranded in the country, for we did drive them to the capital at the end of the day. The agitated physician confessed how distraught he had been at the thought of being abandoned.

The example shows again the inadequate transportation available to a government agency; the cavalier behavior of a higher official who appropriated an assigned automobile without making arrangements for his staff's work to be done (the official himself no doubt had an assignment requiring transportation that would not have been available unless he had requisitioned the health team's car); the immediacy with which the anxious and distrustful response emerges in employees and is directed to supervisors and other personnel, and the lack of appreciation of the effects on employee morale and convenience when an agency fails to provide logistical support.

Example Four: We were preparing to move the examination equipment out of the schoolhouse for transportation down the mountain to the plains village. We asked one of the health-team scribes to help us carry the scale. He became angry and defensive, protesting that he was a "government employee" and would not do manual work. He did not intend to do any work other than that involved in his job. We understood that his status had been threatened, and took time out from our moving to discuss with him what he had been told of his assignment to us in the country. He had been told nothing, it appeared; he was merely ordered to accompany the physicians and nurses and do his job. We spent some time explaining the goals of the study, the purpose of the health examination, and the need for all of us to pitch in to do whatever work was required, relating his own role to the project. We told him how very much he could help if he wished to do so, but that it would be his own choice; that in any event he would be working with us as an equal and we would welcome his suggestions and criticisms. Within 15 minutes a previously sullen, unenergetic white-collar worker became a talkative, immensely helpful, highly constructive member of the team. For the remainder of his time with us on the project he showed initiative, spontaneity, and efficiency, contributing in a variety of ways to the work of the team.

This example shows how important it is to educate employees to the over-all purpose of the work in which they are engaged and to stimulate their personal ego-involvement in that work. As industrial psychologists have shown, this requires that administrators be aware of status problems and of the mechanisms of work-group dynamics. Handled carelessly, the intelligent, sensitive, low-status employee can become a bitter saboteur of his own and other people's work. Handled with understanding, given proper training and the opportunity to gain some personal satisfaction from his job—in a manner that feeds rather than threatens self-esteem—the employee can become an asset, contributing far more in effort and ideas than one could ordinarily demand on the basis of his pay or rank. But to use human resources well, supervisors must be able to abandon traditional authoritarian-destructive techniques and embark on personnel policies which, while they are based on firmness and clear demands for high-level performance, provide the employee with a sense of dignity, worth, intelligent purpose, and reasons for an emotional commitment of interests to his job, associates, organization, and supervisors.

Example Five: Plans for the clinical examination of the villagers called for a mobile x-ray unit. We inquired of health authorities in various public and private agencies, and were given the names of four that had mobile x-ray apparatus. To our surprise, when calling on them, we learned that three had no such apparatus. The fourth did, but officials there explained that their restricted budget prevents them from using the mobile x-ray unit—even though it appears to be the only one in Greece. They had used it once, some six months before on a project financed by the government, but, they reported sadly, the government had not yet paid for the gasoline or film consumed. They hoped to be able to run the unit sometime in the future, when a foreign charity made funds available for gasoline and film. The officials were very cooperative and very interested, but there was no money to be had.

The example demonstrates the presence of a capital investment for which maintenance funds are not budgeted, and shows the effects of the minimal amount of the national budget assigned to health activities. It also shows the surprising lack of information among prominent health authorities about available facilities. This may be ascribed to communication failures, to the very limited time and opportunity for persons in the health field to get together to exchange views and data, and, on a more psychological level, to the restricted scope of attention that occurs in response to the anxiety, frustrations, and heavy work loads burdening most senior management officials in private and public agencies. It has also been suggested, on a more clinical basis, that Greek child-training does not emphasize orientation to time, place, or circumstance, and that many Greeks show an uncommon disregard for or inattentiveness to their surroundings. Insofar as this personality characteristic does exist, it helps to explain some of the startling misinformation that one encounters among supposedly informed personnel.

Example Six: An agency official was directed by his superior to provide medical supplies to a rural facility. The consignment was to be on hand one month later. The requested supplies were unusual and costly, but necessary for the facility's operations. The superior's order had been given in response to this need, and reflected his intelligent appraisal of area conditions. The agency official in turn promised the facility chief that the supplies would be there. The official then went on vacation.

He returned in one month and was asked by the facility chief where his much needed and solemnly promised supplies were. The official was casual about it, denying that he had made any promise of delivery, and freely admitting he had not ordered the supplies. The facility chief reminded him of the order of the official's own superior. The official showed little concern, saying that he had never received an order from the senior administrator of the agency, and that for such unusual supplies the senior administrator himself would have to give the order. The official said that he intended to see the senior administrator within the week and would inquire to see if the ordered supplies would be approved by him. Later in the week the facility chief called the supply official, only to learn that he had not bothered to mention the critical supplies to the senior administrator.

Finally, in desperation, the facility chief himself called on the agency's senior administrator, who agreed that the supplies were necessary and telephoned an order to the supply official. No criticism was made of the latter's failure to abide by the order of his own superior or his failure to be true to his promise.

This example shows how mistrust and anxiety can grow in a responsible authority who tries to perform his task, but who is dependent on others for support or supplies. The intricate pattern of relations within the agency demonstrates the absence of a direct chain of command, and shows how power is concentrated in the agency's senior administrator. Subordinates are reluctant to show any initiative, do not respond to or trust the orders of those immediate superiors whom they recognize as being powerless, and are careful to defend their own position by taking even minor matters to the highest echelons. Authority is not delegated, power is not shared, and in this case a promise served only to relieve an administrator of temporary pressures, but was meaningless to him as a commitment to action. The facility chief, hopelessly dependent in this case, learned that words were meaningless, that orders meant nothing, and that only the wheel of power, its spokes emanating from the senior administrator, was the critical communication pattern.

We turn now from this cursory outline of some of the national problems that impinge on health organizations to a brief discussion of special problems that arise when Greek or foreign health or welfare agencies or groups are engaged in direct efforts to better local conditions.* The task is to bring about technological change in a way that will disrupt the

* For excellent discussions of the approach of technological change and its health implications, the reader is referred to the following publications:

Dorrian Apple, *Sociological Studies of Health and Sickness* (New York, 1960); George M. Foster, *Traditional Cultures and the Impact of Technological Change* (New York, 1962); Edward T. Hall, *The Silent Language* (New York, 1959); Margaret Mead, *Cultural Patterns and Technical Change* (Paris, 1953); Benjamin P. Paul, ed., *Health, Culture, and Community: Case Studies in Public Reactions of Health Programs* (New York, 1955); Lyle Saunders, *Cultural Difference and Medical Care* (New York, 1954); and Edward H. Spicer, ed., *Human Problems in Technological Change* (New York, 1952).

positive sources of community values, morals, and security as little as possible. The task is to perform one's work so that community members remain open and receptive rather than becoming defensive and rigid. In approaching this task, the outsider, whether he is a foreigner or a city Greek, must recognize his special role as a stranger and use the special attributes in the service of his goals which that role implies for the villagers. The goal is usually to create individual and community involvement so that local people can continue with and build upon efforts initiated by outsiders. Most public-health tasks in rural areas require program acceptance by the villagers, program understanding through gentle education, and stimulation of autonomous motivation so that gains will be maintained after the stranger leaves.

To reach these goals, one must fully understand the network of beliefs and social relations within which one is working. One must understand the local culture, the location of power and vested interests in the social structure, and the attitude taken toward government personnel and technological change. One must continually assess one's efforts in order to see how local people perceive the program and its personnel, what beliefs they have about the motives of the staff, and what beliefs they have about the consequences of the program for them. Constant feedback, constant efforts to coordinate staff and villager aims and views, constant care to avoid affront, misunderstanding, false hopes, or entanglement in the snares of local rivalries—these are fundamental requirements. Here are a few examples that speak for themselves:

Example One: A foreign philanthropic agency worked in an area in northern Greece for some time. When the project was to terminate they decided to build, as a final gesture in support of local sanitation, a modern public toilet. They did so, and in a grand ceremony turned the key over to the mayor, admonishing him to take good care of the toilet, putting into practice what they had been teaching about cleanliness. The following year the agency representative returned to the community. He was warmly received; the church bells rang, and the people turned out on the main square. The mayor greeted him happily and within a few minutes proudly told his guest how they had indeed done as he had asked and kept the new toilet clean. Saying this, and with a flourish, the mayor handed the key back to the philanthropist, explaining solemnly that the toilet had been kept absolutely spotless—for he had kept the key and allowed no one to enter the building.

Example Two: Reports that a magician was practicing in an agricultural community filtered into the provincial capital. No one knew where he had come from, but within a few years he had built a great reputation for himself. There was no medical doctor in the town, and all the sick and bewitched were coming to the magician for help. Contrary to law, the magician was handing out various drugs or brews for individuals to take, and was charging fees for each client who came. He was making a good income, it was said.

Under instructions from the capital the magician was arrested. He wanted to explain himself, but no one would listen. The authorities kept him in jail and did not interrogate him beyond the point where they learned he intended to plead "not guilty."

Court opened, and the magician was charged with practicing medicine without a license. Put on the witness stand, the magician, with a mixture of embarrassment and triumph, showed the judge his certificates: a medical diploma and a license to practice. He had, he explained, been unable to attract patients as a professional physician, but presenting himself as a magician and treating his patients under that guise, he had had great success.

Example Three: A health team was setting up for clinical examinations. They prepared the villagers several weeks in advance, explaining, persuading, and stimulating them to cooperate. Team doctors had little time to see everyone, so advanced appointments were made with each family, designating the time they should arrive at the schoolhouse (the examination center). The villagers understood the need for such scheduling, and promised to be on time. When the appointed day came and the doctors arrived at 8 A.M., some villagers had been waiting for hours. Others came in hours late. The villagers could not explain their lack of punctuality until one of the doctors asked if their watches didn't keep time. They replied that no one owned a clock or a watch.

Example Four: During our examinations one patient was found to be suffering from a serious eye disease requiring further examination and treatment in Athens. The social worker told him she would make arrangements for him to be examined, and returned a few days later to his house. He was not at home, so careful directions and the free-pass slip required were given to his wife.

On the follow-up visit the patient was asked if he had gone for his ophthalmological examination. He said he had gone to Athens, but had become lost, did not know which hospital to visit, and had come home unexamined, losing an entire day in the process.

Inquiry revealed that he had no idea that there was more than one hospital or clinic in Athens, and that he did not conceive of the need for any papers or forms once the social worker had told him that arrangements would be made. In addition, his wife had not bothered to tell him of the social worker's call, or of her having left directions and the necessary papers. As an afterthought, the social worker realized that she did not know if either of them could read.

Example Five: At the conclusion of the study one woman upbraided us bitterly. We had come to the village, she said, exploited it, given it nothing. We were not, she said, prepared to provide all medical care, all those medicines, which her family needed. What was worse, we had singled out the neighbor woman she hated worst of all and had given that woman free medicine and had driven her to Athens (a woman acutely ill, probably with tuberculosis), while we had not taken any of her family to the Athens hospital (none were ill although her children were malnourished). Her enemy must have influence with us—what had that woman done to buy us? Her hopes for needed medical care had been dashed, her enemy had triumphed; it would have been better if we had never come to her village at all.

Example Six: At the conclusion of the study, two of the more respected citizens of a village came to us and said: "Now it is time for you to do something more. You know more about us than we do. You are strangers and can tell us what is wrong. If anyone of us tries, they think we want something for ourselves; anyone here who tries to get anything done is distrusted; if he moves ahead they envy him, they eat him up. They think strangers are above this. Look at us; everyone distrusts us, we cannot work together because we are split into factions and still have the old feuds; our officials are uneducated and have no sense of mission or responsibility; they manage things poorly and the tax money goes into their pockets. People don't know how to plan or work together. It isn't enough to find out about our diseases or how to treat them. There is so much more to be done, and everything is tied together."

Voicing the same thought, one of Greece's great physicians, Dr. Spyros Doxiadis, discussing plans for a new program, said:

> Effective work requires an immediate and realistic appraisal of the limitations of our country, of the immediate situation, and of oneself. One cannot set one's goals too high in an underdeveloped country or one either will become totally discouraged and hopeless or will have to leave. Goals must be less high than elsewhere, and satisfactions must come from lesser achieve-

ments. One must remain cheerful—that is an absolute necessity—or one could not continue to live or work.

There can be no marked progress in one area while the rest of the society remains backward; health and education cannot advance unless there is equal simultaneous progress in all other sectors of the economy. It is foolish to expect effort in one area to bring dramatic results, because on all sides the necessary supporting or nourishing conditions which it needs to flourish will not exist. Nevertheless, one should keep on trying, because sometimes an attainment well above the average for that country will provide the stimulus for a general raising of standards in all related fields. There are examples of this, and these are sometimes the only hope for persons who, not being involved in politics, have no other way to contribute to a general improvement of health and education.

Appendixes

Glossary

References Cited

General Bibliography

APPENDIX 1

Methodological Considerations

THE STANDARDIZED QUESTIONNAIRE

We found that the formal questionnaire offers a relatively efficient means for collecting standard information on a number of specific points. It allows one to see what each person—or, in our case, each family—in a given community has to say about a given subject. It is an excellent means for building a catalogue of information about the age, education, size, illness history, and health behavior of these families, as well as about their views on how one should act in given situations, their attitudes toward significant figures in the community, and their knowledge of hygiene. Such information can be used to present a composite picture of the community. This picture is a useful abstraction; it allows one to state how much agreement there is on a given belief or practice, thereby giving indices of homogeneity and heterogeneity within a community.

Questionnaire data also help one to avoid falling into the trap of generalizing from what a particularly vivid and persuasive village informant has to say. Such an informant's views are, in fact, those of a minority, and may be even unique, since his verbal gifts reflect unusual intelligence or creativity. As an illustration of the pitfalls of using the informant system exclusively, one may hear from one that the priest is the moral force in the peasant life and community, and that, by inference, the peasant conducts himself in a way that will earn him the blessing and approval of this revered member of the elite (66). We did indeed meet villagers who expressed reverential sentiments, and who gave evidence of moral conduct not unmindful of the approval or the blessings of the priest and the moral values he represents. Some did use priests as healers. While the presence of such acts and sentiments allows one to say that they are part of the culture, one must have normative data to say how much they are a part of life in a particular village. In this instance in response to a question about what kind of conduct was deserving of the good wishes or blessing of the priest, 52 per cent of the village families did specify moral conduct, *but* 43 per cent rejected the priests outright, calling them "merchants" or worse, and implying that moral conduct was independent of priest or church or abstract religion. Given the latter sentiments, one makes inferences about lack of villager solidarity vis-à-vis the priest, and one is not surprised to find that—in contrast to what the priests contend—few of our villagers ever employ a priest either to bless or to heal them.

One might come to similar conclusions without using a formal questionnaire or making formal visits to every family in the village. But without some approach allowing quantification, one would never be quite so sure about the extent to which the priests are rejected or about the position of one's informant with respect to the rest of the community. It is for reasons such as these that we find the questionnaire useful when applied to community studies.

Questionnaires have disadvantages. Their precision tends to restrict the spontaneity and the scope of conversation in a family visit. Replies to questions may be inexact for a number of reasons—faulty recollection, misunderstandings, the introduction of bias by the interviewer, unwillingness to discuss embarrassing or intimate materials, and so forth. Furthermore, when one is seeking information about health, as we were in the morbidity-survey section of our questionnaire, one can be sure that answers will be affected by a variety of psychological factors. For example, recent studies of morbidity surveys show the following:

(1) The degree of correspondence between individual reports and validating examinations (by physicians in morbidity studies) decreases as the difference between the layman's and the physician's concepts and terminology increases (74). Correspondence between respondent and doctor is highest (but never perfectly matched) when specific, easily identified conditions are involved for which both layman and physician share similar concepts and language (60).

(2) The recency of the occurrence or presence of a condition is associated with the accuracy of its report (65).

(3) The likelihood that a condition or event will be reported by a respondent is associated with how readily he recognizes the condition. The more readily recognized the condition, the more likely there will be recall, other factors being equal (74).

(4) Persons report more accurately for themselves than for others, even when those others are loved ones. Family members may fail to report significant illness experiences involving close relatives, even when those illnesses affect the family (74).

(5) Recall is associated with the intensity of the experience, measured in terms of the time it involves or takes, and the number of associated life consequences. The more intense, the more salient a condition, the more likely it is to be reported (74).

(6) Conditions that are threatening in the sense that their presence arouses anxiety or shame, calls attention to tabooed activities, or concerns acts that are perceived by the respondent as socially unacceptable are likely to be suppressed and thus underreported (74).

(7) Both overreporting and underreporting occur. For ill-defined folk-

conceived conditions, overreporting is common. For carefully defined medical conditions, underreporting is common (60).

(8) Replies given to a trained non-professional interviewer may differ from those given to a professional interviewer. Differences in reporting specific medical conditions may be as high as 62 per cent (60). The social and psychological characteristics of the interviewer in interaction with the respondent shape information given.

(9) Responses are shaped by the respondent's conscious conception of his effect on the interrogator, and of the interactional consequences of his statement on the respondent's own sense of well-being, including his present and future self-interest (60).

(10) The literature of psychiatry and psychology indicates that distortions may be introduced by dynamic factors in personality when matters of emotional or social importance are discussed. Neurotic or parataxic phenomena, efforts to present a social façade or to impress the interviewer, attempts to shape an image of one's self, and reports slanted in the direction of the socially desirable reply are typical distortion processes (19).

(11) In addition to the above, interviewing faces routine problems of interviewer bias, recording error, recall by respondents, failure of the respondent to understand or to have developed opinions or ideas about the item in question, and so forth (34), (36).

To illustrate the kind of error introduced by the method we employed, we call attention to the frequency of reported acute upper-respiratory infections in the family during the preceding 12 months. Twelve months is a long time to remember events, especially events so much a part of life and so temporary as head colds and the flu. Morbidity surveys with larger populations ask about events in a shorter time span; the well-known *Health in California* study (13) asked its respondents to report illnesses experienced during the preceding four weeks, multiplying by 13 to arrive at 12-month incidence rates. One can do this with large samples, but with small populations where the chance that unusual diseases have occurred during the preceding four weeks is 13 times less than the chance that they have occurred during the entire preceding year, one would lose reports of the rare case while gaining recall accuracy.

In California the rate of acute respiratory illnesses per 1,000 persons per year was 1,370/1,000, or about one and one-third cases of cold, flu, bronchitis, etc., per year per person. In our village of Dhadhi the reported rate was 222/1,000, or about one case per every four persons per year. Given the climatic conditions, the life style, and the reports of doctors and villagers emphasizing the frequency of respiratory ailments, we must conclude that there was dramatic underreporting of respiratory ailments and—by implication—of other acute diseases. Using the California rate as a "best guess," one would have underreporting at a rate of one reported case for every five or six acute ailments occurring during the preceding year.

Overreporting probably occurs too. As we have shown in Chapter 4, respondents' reports during the morbidity survey indicated a prevalence rate for heart disease (excluding hypertension) of 27.6/1,000, while medical examination indicated a rate of 9.2/1.000. Other types of overreporting may also be expected in the general interview—i.e., the respondent may give responses that are designed to make a favorable impression on the interviewer. This well-known phenomenon has been studied elsewhere, as for example in the work of Edwards on "response-set" in the direction of social desirability (19).

A different kind of problem arises out of the circumstances of our interviewing. Many of our questions dealt with the family; we interviewed in "natural" circumstances in which several family members might be present during an interview. During any one interview, lasting from one to two hours, from one to seven persons might be present. Family members, neighbors, and children wandered in and out as their curiosity and business dictated. An accurate count of all who participated in any given interview was impossible. Although the head of the household had priority in answering questions, anyone present might reply. Sometimes two or three would answer at once, starting a discussion that might continue until the interviewer interrupted.

We did not wish to restrict spontaneity or to build a "false" interview situation that did not conform to ordinary family conversation modes. Consequently, the interviewer was often faced with a choice of replies, for family members did not always agree on the matter at hand. The general rule was that the answer of the primary respondent as designated by the interviewer—usually the female who stood as head of the household or was spouse to the male head of the house—was given priority in recording.

We could never be sure that the same family members who were present during the first interview would be present during the second, third, or fourth. However, we did require that the primary respondent always be there. One meeting might find her with her children, another with her children and her husband, a third by herself. One can see that a certain inconsistency in replies might well be generated by these varying family groups.

Quite by accident, we were able to make a small reliability check—to measure how much agreement there was on the same questions by different family members. One of our social workers began to interview the male head of a house when it appeared that his wife would not be available. The next day our second social worker, not realizing that the first interview had been conducted, found the wife and began to interview her. The husband had not bothered to tell his wife of his interview the preceding day, a not unusual situation in rural Greece where husband and wife often fail to tell one another very much. As a result, the wife was unaware of what the husband had said. We had hoped to continue in this fashion with this couple, both interviewers completing separate interviews on their respective respondents. Time pressures toward the end of the study forbade that, so we were forced to make our comparisons on partial data only.

The husband and wife answered 15 major questions independently of one another. On eight questions (53%) their replies were identical. On another four questions (27%) they were generally similar and were coded in the same content categories.* On three other questions (20%) they were dissimilar but nevertheless contained elements of overlap or similarity. No replies were totally dissimilar or lacking in common elements. These comparisons provided us with a reliability rate of 80 per cent between husband and wife in one family.

It is of interest to see the kinds of disagreements that arose. For example, take the question "Have you or anyone in this household been sick during the last 12 months?" and then, in the case of an affirmative reply, the following sub-questions: "Who has been sick?" "What illness or illnesses did each have?" "How long was each person sick with each illness?" "How many days were spent in bed for each person and each illness?" "What do you think caused each person's (listed) illness (each mentioned)?" and "What was done for each illness for each person?" (listed by person and requiring probing by the interviewer for actions taken).

In response to this group of questions, the husband said that six family members had been ill, listing as illnesses one kidney disorder, four cases of the flu, pain in the hand from an old accident and operation, and pain in the limbs from walking. The wife listed only her own kidney disorder.

In regard to the commonly acknowledged kidney disorder, the husband said his wife had had the illness (characterized as kidney pains) for one year, that she had spent three to four days in bed occasionally during the past year, that he believed she had been to a local doctor and to an Athenian doctor for treatment, and that he considered the illness cause to be fatigue. The wife, on the other hand, said she had been sick for three years, that it had caused her to go to bed off and on for periods of three to four days during the last year, that she had treated herself with various herbs known in the local lore, but that she had *not* been to the doctor and that the cause was to be attributed to her lifting heavy objects and to fatigue.

In response to another question, "During the last 12 months has any member of the household used any plants, herbs, drugs, special waters, or any other medicine or thing in order to protect their health or to help cure any bodily ailment or bodily trouble?" both husband and wife agreed that healing measures had been taken, but the wife could not list any specific measures employed, even though she had earlier mentioned herbs, while the husband recalled that holy water and herb teas had been used in treatment.

In reviewing these examples, one sees that disagreement within the family can occur on matters of considerable importance to the family, and to us in our study. The wife mentioned above may have a different definition of what

* Content coding is the method by which interview replies were grouped into categories, a method that allowed us to summarize the replies and draw comparisons among respondents.

constitutes sickness than does her husband; but it may also be that she is more self-centered, that she takes no interest in her husband and his complaints, or that her memory is poorer. In any event, we see the difficulties involved in getting people to report accurately on the ailments of other family members. This finding is also reported by Simmons and Bryant (65).

THE HYGIENE SCALE FOR FAMILIES

One formal assessment device was the brief form of the Hygiene Scale for Families, adopted from its use in Syria by Dodd (18) and modified for local conditions. It consisted of 16 items that required the interviewer either to ask the respondent about or to rate family practices on hygiene, or both. Findings on this scale are subject to several errors, for, as we indicated in Chapter 8, we suspect an implied exaggeration of the protein diet. For example, 14 of 24 families in Dhadhi and seven of 13 families in Panorio declared that they had eaten meat, fish, eggs, or nuts during the preceding 48-hour period. This suggests a dietary balance not consistent with our observations made at mealtime. Allbaugh's (2) report on Crete, where villagers appear to have a better diet than in many Greek areas, finds that only 4 per cent of the caloric intake was accounted for by these foods, as was 18 per cent of the daily protein intake. The problem here involves (a) the phrasing of the question; it does not ask how much meat, fish, etc. was consumed; (b) the method; unlike the method employed in the excellent Crete survey, our method did not include weighing the daily intake per family on a gram scale; and (c) the possible false reporting; the midwife (see Chapter 10) said she believed the villagers were overreporting their meat-fish-eggs-nut menu.

Another problem in the brief form of the Hygiene Scale was the item on milk boiling. No Dhadhi family and only three Panorio families said they did not boil their milk. Observations and spontaneous remarks on other occasions, e.g., "Don't you love the taste of milk foaming fresh from the goats?" indicate a bias here in the direction of the respondents' saying what they knew they ought to do rather than admitting their lapses in daily practice. For families with only wood fires, as Allbaugh observed in Crete, the boiling of all milk presents great difficulty. For the shepherds who are tending the flocks, the temptation to drink fresh milk is great; yet none of our Saracatzani admitted that they did not boil it.

THE MEDICAL EXAMINATION

During the course of the medical examinations, an American-trained physician visited the project and volunteered to examine two ill patients who had presented themselves in the morning before the public-health team had arrived on the scene. After his examination, limited by his having no equipment at all, these same two patients were examined by team doctors. We thus have two cases on which to compare evaluations made by a visiting American-trained doctor with those made by the two Greek university-trained public-health doctors.

There were consistent differences. The American-trained physician reported system findings in more detail, took greater notice of pathology, arrived at diagnoses in each case that suggested possible serious illness, and recommended specialist examinations. As one observed them both during examinations, it was clear that the public-health doctor took much less time (an average of 15 minutes per patient) than the visiting physician (an average of 45 minutes per patient). Although we cannot be sure of the validity of either finding, especially since neither patient cooperated by going to Athens to a specialist, one presumes that the longer examination was more likely to be the accurate one. Given this assumption, it would appear that the public-health physicians underreported illness that under ideal circumstances would have been detected.

Nevertheless, we face a cultural as well as a medical-methodological problem here. American physicians have been charged with overemphasizing pathology, and the accusation is supported by research evidence (5). Greek physicians, on the other hand, may view normal (statistically speaking) dysfunction or tissue "pathology" as part of ordinary life, thereby sharing the perspective of the peasant, who expects to experience and cohabit with a host of troubles. Such a point of view is in contrast with that held by the American—whether patient or doctor—who expects nature to submit to control, and who views sickness or trouble as an alien intruder to be vanquished immediately.

Considering the speed with which it was necessary for the public-health team to conduct examinations, real underreporting would not be surprising. Our observations of the team doctors revealed that in their examinations of most patients they hurried through the history, did not completely undress any female and only a very few male patients, did not examine the conjunctiva, did not do a visual field examination, did not examine eye grounds or ears, did not always inspect the mouth or throat, did not always examine the neck, rarely examined all of the lymph-node areas (neck, armpit, groin), did not always examine the abdomen, rarely checked limbs, never did a rectal or vaginal examination, and never examined the breasts. Thus it is unlikely that the results of the public-health–team examination, valuable as they were, fully reflect the morbidity status of the villagers.

A Western-trained medical specialist observed our clinical examination and said:

The medical work observed does not differ from much of the typical practice of an American general practioner. It does differ from that of a highly trained specialist in internal medicine. For example, the internist is trained to consider that the most important part of a physical examination is a good history. During the observed examination only a sketchy history was taken. The observed physician looked first at the area where the patient said his difficulty was: the throat and neck. A general practitioner would have done the same. An internist would have looked at that area later. The physician was brusque and curt, abrupt and aggressive; he did not approach the patient gently, but pushed and pulled him about. He looked in the throat, using a tongue depressor and the light from the window despite the fact that a flashlight was available. He did not look in the ears, although in a case of naso-laryngitis, as presumed here, minimum standards of practice would require an examination of the ear-drum. He went over the heart and chest very quickly with a stethoscope, but he did not do percussion or other tactile examination of the chest wall. He did not examine the abdomen, nor did he look for tenderness in the kidney area. He did not do a neurological examination in spite of the patient's history of meningitis. In all probability the patient had a variant of the common cold and the doctor had enough experience to recognize it as such. Therefore the doctor looked for the obvious. He certainly did not look for the unexpected or the unlikely. Chances are the doctor arrived at the correct diagnosis for the wrong reasons.

The examination could not be termed a screening physical, for it was focused only on the area of complaint. It was, by specialists' standards, a poor examination, but many physicians practicing all over the world would have done it in just the same way.

Assuming that the morbidity study did underreport illness, we cannot attribute the resulting errors only to haste, inadequate medical training, or the lack of laboratory and other specialized equipment. More pervasive cultural factors also played a role.

By cultural factors we mean, for example, that women patients were unwilling to undress before the physician. Many had to be assured when their appointments were made that they would not be required to disrobe. While modesty of the women plays a strong role here, one cannot overlook the attitude of their men, who were said to be vehemently opposed to their wives' and daughters' undressing before other men. The role of physician in this instance provides no freedom from the ordinary restraints or taboos, and thus the physician as a man is prevented from conducting himself as his medical role requires. Greek physicians were quick to recognize this, expressing fear for their lives should they undertake to ask a peasant woman patient to undress.

Such strictures of modesty and of male possessiveness pose serious problems in the detection and treatment of disease. For example, one day during our study the public-health nurse was doing a chest measurement of a peasant woman. The nurse noticed that a nipple was inverted and asked the patient about it. The patient said she was aware of the condition, adding that her breast was sometimes sore and that she had felt a hard mass within it. The nurse, alarmed, instructed the patient to be sure to tell the physician about this condition. Several days later, the nurse remembered to ask the physician what he had diagnosed in the woman. The doctor was surprised, for no woman had mentioned anything to him about a breast problem, and he, of course, had not done any breast examinations. The nurse tried to recall the patient's name or physical description, but could not do so. The team did its best to discover which patient it might have been, but unfortunately they were unsuccessful. If the patient had breast cancer, as suspected, a fatal outcome must be anticipated in a case that might well have responded to treatment if only the modesty of the woman had not restrained her from giving relevant information to the physician. In any event, we can be sure that considerable underreporting of diseases of the breast, the rectum, and the male and female sexual organs has occurred.

APPENDIX II

Findings and Recommendations

1. *Body System Findings*

System Signs and Diagnosis	Number of Cases Dhadhi (N = 144)	 Panorio (N = 91)	 Saracatzani (N = 37)
Skin and appendages			
Eczema	3		1
Jaundiced color	1		
Burn	1		
Inflamed		1	
Allergy		1	
Psoriasis		1	
Lymphatic and Hematic			
Nephritis	1		
Scrofula		1	
Inguinal swelling		1	
Head, Face, Neck			
Glands palpable	2		1
Hairless, abnormal	1		
Multiple lymph-gland enlargement	1	2	
Swollen face		1	
Nose, Sinuses, Mouth, Throat			
Sensitivity in tonsils			1
Overgrowth in tonsils, swollen tonsils	12	3	3
Tonsillitis	5	2	2
Sensitivity of nose, pharynx			4
Rhinitis, pharyngitis	3	4	1
Quinsy, or acute sore throat	8	4	1
Nasal speech		1	1
Quinsy, upper respiratory infections associated with shoulder pain	1		

System Signs and Diagnosis	Number of Cases Dhadhi (N = 144)	Panorio (N = 91)	Saracatzani (N = 37)
Ears			
Hearing: good	90	67	28
Hearing: fair	25	23	4
Hearing: poor	6		
Ringing in ears (tinitis)	5	1	2
Pain, ache	2		
Otitis	2		
Dizziness associated with hearing loss	1		
Eczema		1	
Eyes			
Visual acuity: good	92	67	25
Visual acuity: fair	21	16	4
Visual acuity: poor	8	5	5
Astigmatic			1
Blind		1	
Cataract	1	1	
Cardiovascular			
High blood pressure	9	3	1
Tachycardia	2	2	2
Varicose veins	5	3	
Respiratory			
Asthmatic bronchitis	2		
Dystrophic chest	2		
Malformed chest	1		
Chronic bronchitis	1		
Digestive system			
Indigestion	14	4	4
Gastritis	1		
Pain above belly (otherwise unspecified)	2	3	
Lack of appetite	1	1	
Gastric upset associated with menstruation	1		
Ulcers	2	1	
Abdominal pain	2	1	1
Anorexia	1	1	
"Heaviness"	1		
Diarrhea and constipation		1	
Hyperhydrochloric		1	
Hernia			
Operated	1		
Inguinal hernia, rupture	7	6	
Umbilical hernia	2		

System Signs and Diagnosis	Number of Cases Dhadhi (N = 144)	Panorio (N = 91)	Saracatzani (N = 37)
Genito-urinary			
Menstrual difficulty	1		1
Overfrequent urination	4		
Loss of urinary control	2		
Disorder connected with menopause	1		
Difficulty in urination	2		
Operated for prostate, bladder	1	1	
Pain in kidney area after exposure to cold		1	
Hysterectomy	1		
Musculo-skeletal			
Fracture (recent)			1
Rheumatic pain	2		
Gout	1		
Infected leg	1	1	
Limitation of motion: rheumatic-arthritic	2		
Thumb with double phalanx	1		
Hemiplegia		2	
Hands; fingers: handicapped		2	
Burns		1	
Wound		1	
Post-fracture limitation of motion: elbow		1	
Endocrine	—	—	—
Nervous system			
Hypotonic reflexes	5	2	4
Hypertonic reflexes	14	6	2
Headache	2	1	
Autonomic disturbance	2		
Psychiatric			
Mental retardation		1	

2. *Medical Recommendations Given to Patients*

Recommendations	Dhadhi (N = 144)	Panorio (N = 91)	Saracatzani (N = 37)
General			
Take medication (given free by medical team or to be bought as prescribed)	37	26	10
Follow special restrictive diet	25	3	4
Get more food, better nourishment, vitamins	20	13	4
Take sea baths, sand baths	15	3	5
Better self care, more rest	10	8	0
Continue treatments begun by other (previous) doctor	7	2	2
Use hernia belt, prosthetic device	3	1	1
Use gargle, rinse, douche (no drug prescribed)	1	0	3
Undergo bloodletting	1	0	0
Restrict alcohol consumption	1	1	1
Other	6	2	3
Total persons receiving general recommendations	90	68	28
Special			
See a specialist	25	13	5
Have special laboratory tests	6	14	0
Have tests and see a specialist	2	4	2
Total persons receiving special recommendations	33	31	7
Total persons receiving general and special recommendations	105 (73%)	68 (75%)	28 (76%)

Glossary

Aerika. Air spirits.
Anemopyroma. Facial erysipelas.
Apotropaic rites. Rituals of avoidance, to ward off pollution, evil spirits, or danger.
Bad Hour. Spirits or demons, forming a subgroup of the exotika, which usually take animal forms; thought to be a cause of illness, especially in children.
Bad pimple. Anthrax.
Bubble, the. Folk ailments.
Charon. Personification of death; ferryman of the River Styx in Hades.
Chthonic deities. Gods of the underworld.
Cutting the jaundice. Cutting the labial frenum, a membrane connecting the upper lip to the gum; a supposed cure for jaundice.
Efchelaio, Efchi. Special prayers and blessings.
Elafroiskioti. "Light-shadowed" persons, reputedly gifted with second sight and able to see exotika.
Exotika. Broad class of supernaturals outside the Christian religious system.
Ex votos. Thank-offerings for successful healing, usually shaped like the limb or organ cured.
Heaviness. Folk ailment, possibly of the digestive system.
Hybris. Overweening pride.
Iatressa. Wise woman specializing in herbs.
Kallikantzari. Hairy people, a subgroup of the exotika.
Ker, the (Keres). Spirit of disease.
Komboiannitis. Folkhealer specializing in bone and joint cases.
Korakiasma. Folk ailment.
Koumbaros. Godfather, also best man at a wedding.

Lechones. Postpartum mothers.
Light-shadowed. *See* Elafroiskioti.
Messa. Influence.
Moirai. The fates.
Moros. Subgroup of the exotica.
Neraides. Water nymphs.
Panaghia. Virgin Mary, the "all holy."
Pharmakos. Scapegoat.
Philotimo. Self-esteem, honor.
Practika. Traditional folk remedies.
Practikoi. Komboiannites, folkhealers specializing in bone and joint cases.
Rites of passage. Ceremonies that mark important occasions such as birth, the onset of puberty, or marriage.
Sorcery. Deliberate, malicious magic.
Soul child. Adopted child or adult.
Therapea (*Tendance*). Service to supernaturals given in the hope of receiving benefits in return.
Vittora. Evil spirits.
Vrikolakes. Revenants, spirits of the dead unable to find rest.
Waist out of place. Folk ailment, backache.
Wandering navel. Folk ailment.
White spot in the eye. Folk ailment.
Witchcraft. Use of magic without conscious intent to harm.
Xemetrima. Ritual words used in healing.

References Cited

1. Ackerknecht, E. W. "The role of medical history in medical education," *Bull. Hist. Med.*, 1947, **21**, 135–45.
2. Allbaugh, L. Crete: A Case Study of an Underdeveloped Area. Princeton, N.J.: Princeton Univ. Press, 1953.
3. Apple, D. Sociological Studies of Health and Sickness. New York: McGraw-Hill, 1960.
4. Ashley Montagu, M. F. "Primitive medicine." *New England J. Med.*, 1946, **235**, 43–49.
5. Bakwin, H. "Pseudodoxia pediatrica," *New England J. Med.*, 1945, **232**, 691–97.
6. Bard, M., and Ruth B. "The psychodynamic significance of beliefs regarding the cause of serious illness," *Psych. Rev.*, 1956, **43**, 146–62.
7. Benedict, Ruth. Patterns of Culture. New York: Mentor Books, 1950.
8. Biris, K. Arvanites: The Dorians of Modern Hellenism. Athens, 1960 (in Greek, publisher not cited).
9. Blum, Eva. "The Uncooperative Patient," in *Supplementary Studies on Malpractice*. San Francisco: Calif. Med. Assn., 1958.
10. Blum, R. H. The Management of the Doctor-Patient Relationship. New York: McGraw-Hill (Blakiston), 1960.
11. Blum, R. H. The Psychology of Malpractice Suits. San Francisco: Calif. Med. Assn., 1957.
12. Bury, J. B. A History of Greece to the Death of Alexander the Great. New York: Modern Library, 1925.
13. Calif. Dept. of Public Health. Health in California. 1956.
14. Cartwright, Ann. "Some problems in the collection and analysis of morbidity data obtained from sample surveys," *Milbank Mem. Quart.*, 1959, **37**, 33–48.
15. Clark, Margaret. Health in the Mexican-American Culture. Berkeley: Univ. of Calif. Press, 1959.

16. Cumont, F. Astrology and Religion among the Greeks and Romans. New York: Dover Publications, 1960.
17. Dakaris, S. "The dark palace of Hades," *Archaeology*, 1962, **15**, 85–93.
18. Dodd, S. "A controlled experiment on rural hygiene in Syria," American Univ. of Beirut, Social Science Series No. 7, 1934.
19. Edwards, A. The Social-Desirability Variable in Personality Assessment and Research. New York: Dryden, 1957.
20. Elworthy, F. T. The Evil Eye. New York: Julian Press, 1958.
21. Fontenrose, J. Python: A Study of Delphic Myth and Its Origins. Berkeley: Univ. of Calif. Press, 1959.
22. Foster, G. M. Traditional Cultures: The Impact of Technological Change. New York: Harper, 1962.
23. Francis, E. K. L. "The personality type of the peasant according to Hesiod's Works and Days: A cultural case study," *Rural Sociology*, 1945, **10**, 275–95.
24. Frank, J. D. Persuasion and Healing. Baltimore: Johns Hopkins Press, 1961.
25. Frejos, P. "Man, Magic and Medicine," in I. Galdston, ed., *Medicine and Anthropology*. New York: Int. Universities Press, 1959.
26. Friedl, Ernestine. "Hospital care in provincial Greece," *Human Organization*, 1958, **16**, 24–27.
27. Friedl, Ernestine. Vasilika: A Village in Modern Greece. New York: Holt, 1962.
28. Garnett, Lucy, and J. S. Stuart-Glennie. Greek Folk Poesy. London: David Nutt, 1896.
29. Harrison, Jane E. Epilegomena to the Study of Greek Religion and Themis: A Study of the Social Origins of Greek Religion. New Hyde Park, N.Y.: University Books, 1962.
30. Harrison, Jane E. Prolegomena to the Study of Greek Religion. New York: Noonday, 1955.
31. Hatzimichalis, A. Sarakatzani. Athens: Kakoulithis, 1957. 2 vols.
32. Hesiod. The Works and Days, Theogony, The Shield of Herakles. Translated by Richmond Lattimore. Ann Arbor: Univ. of Michigan Press, 1959.
33. Homer, The Iliad. Translated by Richmond Lattimore. Chicago: Univ. of Chicago Press, 1951.
34. Hyman, H., *et al.* Interviewing in Social Research. Chicago: Univ. of Chicago Press, 1954.
35. Institute for Thoracic Diseases, Athens. Personal communication, 1962.
36. Kahn, R. L., and C. F. Cannell. The Dynamics of Interviewing. New York: Basic Books, 1958.
37. Kemp, P. Healing Ritual: The Technique and Tradition of the Southern Slavs. London: Faber & Faber, 1935.
38. Koos, E. The Health of Regionville. New York: Columbia Univ. Press, 1954.

39. Lawson, J. C. Modern Greek Folklore and Ancient Greek Religion. Cambridge, Eng.: Cambridge Univ. Press, 1910.
40. Lee, Dorothy. "Greece," in M. Mead, ed., *Cultural Patterns and Technical Change.* Paris: UNESCO, 1953.
41. McNeill, W. Greece: American Aid in Action 1947–1956. New York: Twentieth Century Fund, 1957.
42. Megas, G. A. Greek Calendar Customs. Athens: Press and Information Dept., 1958.
43. Merton, R., and Alice Kitt. "Contributions to the Theory of Reference Group Behavior," in R. K. Merton and P. F. Lazarsfeld, eds., *Continuities in Social Research: Studies in the Scope and Method of "The American Soldier."* Glencoe, Ill.: Free Press, 1950.
44. Morris, J. N. Uses of Epidemiology. Edinburgh: Livingston, 1957.
45. Naroll, R. "Controlling data quality," Symposia Study Series No. 4, Natl. Inst. of Social and Behavioral Science, Series Research in Social Psychology, Washington, D. C., 1960.
46. Nietzsche, F. The Birth of Tragedy. New York: Modern Library, 1954.
47. Nilsson, M. P. A History of Greek Religion, 2d ed. Oxford: Clarendon Press, 1952.
48. Nilsson, M. P. The Minoan-Mycenaean Religion and Its Survival in Greek Religion. Lund, Sweden: Gleerup, 1950.
49. Papandreou, A. Personal communication. 1962.
50. Peterson, O., *et al.* "An analytical study of North Carolina general practice," *J. Med. Ed.*, Pt. II, 1956, **31**.
51. Pitt-Rivers, J. A. The People of the Sierra. London: Weidenfeld & Nicolson, 1954.
52. Plato. Gorgias. Revised, with Introduction and Commentary by E. R. Dodds. Oxford: Clarendon Press, 1959.
53. Politis, N. G. Paradosis. Athens: Sakelariou, 1904.
54. Radin, P. Primitive Religion. New York: Dover Publications, 1957.
55. Redfield, R. The Little Community. Chicago: Univ. of Chicago Press, 1955.
56. Redfield, R. Peasant Society and Culture. Chicago: Univ. of Chicago Press, 1956.
57. Rogler, L., and A. B. Hollingshead. "The Puerto Rican spiritualist as a psychiatrist," *Am. J. Soc.*, 1961, **67**, 17–21.
58. Rose, H. J. Gods and Heroes of the Greeks. New York: Meridian Books, 1958.
59. Rose, H. J. Religion in Greece and Rome. New York: Harper, 1959.
60. Sanders, B. S. "Have morbidity surveys been oversold?," *Am. J. Pub. Health,* 1962, **52**, 1648–59.
61. Sanders, I. Rainbow in the Rock: The People of Rural Greece. Cambridge, Mass.: Harvard Univ. Press, 1962.
62. Sharpe, J. C., and W. L. Marxer. "Physical examination of well persons," *Calif. Med.*, 1962, **96**, 35–40.

63. Sigerist, H. On the Sociology of Medicine. Edited by M. Roemer. New York: MD Publications, 1960.
64. Sigerist, H. "The Physician's Profession Through the Ages," in Félix Martí-Ibáñez, ed., *On the History of Medicine*. New York: MD Publications, 1960.
65. Simmons, W. R., and E. E. Bryant. "An evaluation of hospitalization data from the health interview survey," *Am. J. Pub. Health*, 1962, **52**, 1638–47.
66. Thomas, W. I., and F. Znaniecki. The Polish Peasant in Europe and America. New York: Dover Press (reissue), 1958. 2 vols.
67. Tomasic, D. "Personality development of the Dinaric warriors," *Psychiatry*, 1945, **8**, 449–93.
68. Tomasic, D. "The Structure of Balkan Society," in R. Bendix and S. M. Lipset, eds., *Class, Status, and Power*. Glencoe, Ill.: Free Press, 1953.
69. Trussell, R. E., and J. Elison. Chronic Illness in a Rural Area. Cambridge, Mass.: Harvard Univ. Press, 1959.
70. U. S. Public Health Service. Medical Care Financing and Utilization. Health Economics Series No. 1. Washington, D.C.: Govt. Printing Office, 1962.
71. Vasilios, G. Man's Hygiene. Athens, 1962.
72. Weinberg, S. K. Incest Behavior. New York: Citadel, 1955.
73. Williams, Phyllis H. South Italian Folkways in Europe and America. New Haven, Conn.: Yale Univ. Press, 1938.
74. Woolsey, T. D., P. S. Lawrence, and Eve Balamuth. "An evaluation of chronic disease prevalence data from the health interview survey," *Am. J. Pub. Health*, 1962, **52**, 1631–37.
75. Zborowski, M. "Cultural components in response to pain," *J. Social Issues*, 1952, **8**, 16–30.

General Bibliography

Ackerknecht, E. W. "Primitive medicine and culture pattern," *Bull. Hist. Med.*, 1942, **11**, 503–21.

Adams, R. N. "Notes on the application of anthropology," *Human Organization*, 1953, **12**, 10–14.

Akoglous, X. Laographika Kotyoron. Athens: Xenou, 1939.

Allen, H. B. Rural Reconstruction in Action: Experience in the Near and Middle East. Ithaca, N.Y.: Cornell University Press, 1953.

Angel, J. L. "Human biology, health, and history in Greece from first settlement until now," *Am. Philosophical Soc. Year Book 1954*, pp. 168–72.

Argenti, P. P. Bibliography of Chios. Oxford: Clarendon Press, 1940.

Argenti, P. P., and H. J. Rose. The Folklore of Chios. Cambridge, Eng.: Cambridge Univ. Press, 1949.

Blum, R. H. "Case identification in psychiatric epidemiology: methods and problems," *Milbank Mem. Quart.*, 1962, **40**, 253–88.

Blumner, H. The Home Life of the Ancient Greeks. London: Cassell, 1895.

Bonser, W. "Animal skins in magic and medicine," *Folklore*, 1962, **73**, 128–29.

Buchholz, E. A. W. Die Homerischen Realien; Das Privatleben. Leipzig: Teubner, 1881.

Butler, E. M. Ritual Magic. New York: Noonday, 1959.

Cameron, A. "Folklore as a medical problem among Arab refugees," *The Practitioner*, 1960, **185**, 347–53.

Castiglioni, A. Adventures of the Mind. New York: Knopf, 1946.

Caudill, W. "Applied anthropology in medicine," in A. L. Kroeber, ed., *Anthropology Today*. Chicago: Univ. of Chicago Press, 1953.

———. "Around-the-clock patient care in Japanese psychiatric hospitals: The role of the Tsukisoi," *Am. Soc. Rev.*, 1951, **26**, 204–14.

Childe, V. G. New Light on the Most Ancient East. New York: Grove Press, 1957.

Clark, Margaret. "The social functions of Mexican American medical beliefs," *California Health*, 1959, **16** (May), 1953–56.

Clements, F. E. "Primitive concepts of disease," *Univ. of Calif. Pub. in Am. Archaeology and Ethnology*, 1932, **32**, 185–253.

Crawley, E., and T. Besterman. The Mystic Rose. London: Methuen, 1927. Vols. I and II.

Dawson, W. R. The Bridle of Pegasus: Studies in Magic, Mythology, and Folklore. London: Methuen, 1930.

Delatte, A. Herbarius: Recherche sur le cérémonial usité chez les anciens pour la cueillette des simples et des plantes magiques. Paris: Droz, 1938.

De Vos, G., and H. Wagatsuma. "Psycho-cultural significance of concern over death and illness among rural Japanese," *Int. J. Soc. Psychiat.*, 1959, **4**, 5–19.

De Waele, F. J. M. The Magic Staff or Rod in Graeco-Italian Antiquity. The Hague: Aloysius College, 1927.

Dioscorides, P. Greek Herbal of Dioscorides. Edited by R. T. Gunther. London: Oxford Univ. Press, 1934.

Dodds, E. R. The Greeks and the Irrational. Berkeley: Univ. of Calif. Press, 1951.

Durrell, L. Prospero's Cell, and Reflections on a Marine Venus. New York: Dutton, 1960.

Evans-Pritchard, E. E. The Nuer. Oxford: Clarendon Press, 1950.

———. Witchcraft, Oracles, and Magic among the Azande. Oxford: Clarendon Press, 1937.

Fehmy, I. B. "De quelques usages de l'Anatolie," *Ann. et bull. de l'hôpital d'enfants Hamidié*, 1905, **4**, 47–48.

Fermor, P. Mani. New York: Harper, 1958.

Foster, G. M., ed. A Cross-Cultural Analysis of a Technical Aid Program. Washington, D.C.: Smithsonian Institution, July 25, 1951.

———. Problems in Intercultural Health. New York Social Science Res. Council, pamphlet 12, 1958.

———. "Relationships between theoretical and applied anthropology; a public health program analysis," *Human Organization*, 1952, **2**, 5–16.

———. "Use of anthropological methods and data in planning and operation," *Public Health Reports*, 1953, **68**, 841–57.

Frake, C. O. "The diagnosis of disease among the Subanun of Mindanao," *Am. Anthropologist*, 1961, **62**, 113–32.

Frazer, J. The Golden Bough: A Study in Magic and Religion. New York: Macmillan, 1951. Vols. I–XII.

Fried, J. "Acculturation and Mental Health among Indian Migrants in Peru," in M. Opler, ed., *Culture and Mental Health*. New York: Macmillan, 1959.

Friedl, Ernestine. "The role of kinship in the transmission of national culture to rural villages in mainland Greece," *Am. Anthropologist*, 1959, **61**, 30–38.

Galdston, I. The Meaning of Social Medicine. Cambridge, Mass.: Harvard Univ. Press, for the Commonwealth Fund, 1954.

Galdston, I., ed. Medicine and Anthropology. New York: Int. Universities Press, 1959.

Gelzer, H. Geistliches und Weltliches aus dem türkisch-griechischen Orient. Leipzig: Teubner, 1900.
Glock, C. Y. "The Sociology of Religion," in R. K. Merton, L. Broom, and L. S. Cottrell, eds., *Sociology Today*. New York: Basic Books, 1959.
Gorer, G. (with Dorothy Lee) "The Greek Community and the Greek Child from the Viewpoint of Relief and Rehabilitation." Mimeographed.
Gray, P. People of Poros. New York: McGraw-Hill, 1942.
Guthrie, W. K. C. The Greeks and Their Gods. Boston: Beacon, 1950.
Hall, E. T. The Silent Language. New York: Doubleday, 1959.
Hallowell, A. I. "Culture, Personality and Society," in A. L. Kroeber, ed., *Anthropology Today*. Chicago: Univ. of Chicago Press, 1953.
Hovorka, O. von, and A. Kronfeld. Vergleichende Volkmedizin. Stuttgart: Strecker and Schröder, 1908–1909. Vols. I and II.
Hippocrates. The Genuine Works. Translated by Francis Adams. Baltimore: Williams & Wilkins, 1939.
———. Sämtliche Werke. Edited by R. Fuchs. München: Lüneburg, 1895.
Hueppe, F. Zur Rassen und Socialhygiene der Griechen. Wiesbaden: Kreidel, 1897.
Hughes, C. C. "The patterning of recent cultural change in a Siberian Eskimo village," in R. J. Smith, ed., "Culture change and the small community," *J. Soc. Issues*, 1958, **14**, 27–37.
Jaco, E. Patients, Physicians, and Illness. Glencoe, Ill.: Free Press, 1958.
James, E. O. Prehistoric Religion. New York: Praeger, 1957.
———. Seasonal Feasts and Festivals. London: Thames & Hudson, 1961.
Jayne, W. A. The Healing Gods of Ancient Civilizations. New Hyde Park, N.Y.: University Books, 1962.
Jung, C. G., and C. Kerényi. Essays on a Science of Mythology: The Myth of the Divine Child and the Mysteries of Eleusis. New York: Bollingen Series, Pantheon Books, 1949.
Kyriakides, S. Neugriechische Volkskunde, Volksdichtung, Volksglaube, Volkskunst. Thessalonika, 1936.
Lackman, R., and W. J. Bonk. "Behavior and beliefs during the recent volcanic eruption at Kapho, Hawaii," *Science*, 1960, **131**, 1095–96.
Lambert, W. W., L. M. Triandis, and Margery Wolf. "Some correlates of beliefs in the malevolence and benevolence of supernatural beings: A cross-societal study," *J. Abnorm. Soc. Psychol.*, 1959, **58**, 162–69.
Leographia. "Popular Medicine," *Leographia Archives*, Academy of Athens, Vols. I–XVII, 1909–58.
Levy, H. L. "Property distribution by lot in present-day Greece," *Trans. Am. Philological Assn.*, 1956, **87**, 42–46.
Lewis, O. Life in a Mexican Village: Tepoztlan Restudied. Urbana: Univ. of Illinois Press, 1951.
Liungman, W. Traditionswanderungen Euphrat-Rhein, *FF Communications* (Helsinki: Suomalainen Tiedeakatemia), No. 119, 1938.
Lucian. Alexander, the False Prophet. London: Heinemann, 1925.

Maas, P. L., and J. H. Oliver. "An ancient poem of the duties of a physician," *Bull. Hist. Med.*, 1939, **7**, 315–23.

Magnus, H. Der Aberglaube in der Medizin. Breslau: Kern, 1903.

Mandelbaum, D. G. "Social Uses of Funeral Rites," in H. Feifel, ed., *The Meaning of Death.* New York: McGraw-Hill (Blakiston), 1959.

Marcuse, J. "Die Lehre von der Lungenschwindsucht im Altertum," *Zeitschr. f. diätische und physikalische Therapie* (Leipzig: Thieme, 1899), **3**, Heft 2.

Marketos, B. J., ed., A Proverb for It: 1510 Greek Sayings. New York: New World Pub., 1945.

McDermott, W., *et al.* "Introducing modern medicine in a Navajo community," *Science*, 1960, **131**, 197–205, 280–87.

McDougall, J. B. "Tuberculosis in Greece," *Bull. of the World Health Organization*, 1947–48, **1**, 103–96.

McKenzie, D. The Infancy of Medicine: An Inquiry into the Influence of Folklore upon the Evolution of Scientific Medicine. London: Macmillan, 1927.

McNeill, W. H. Greece: American Aid in Action. New York: Twentieth Century Fund, 1957.

Mavrogordato, J. Modern Greece, Chronicle and Survey. London: Macmillan, 1931.

Ménard, L. Hermes Trismégiste. Traduction et étude sur l'origine des libres hermétiques. Paris: Didier, 1867.

Messing, S. D. "Group Therapy and Social Status in the Zar Cult of Ethiopia," in M. Opler, ed., *Culture and Mental Health.* New York: Macmillan, 1959.

Mireaux, E. Daily Life in the Time of Homer. New York: Macmillan, 1959.

Murray, G. Five Stages of Greek Religion. New York: Doubleday, 1955.

———. The Literature of Ancient Greece. Chicago: Univ. of Chicago Press, 1956. 3d ed.

Near East Foundation, "Annual Report of the Educational Director," 1949 and 1950 (two mimeographed pamphlets for each year).

Near East Foundation, "Near East Relief Consummated; Near East Foundation Carries On." A Committee of Trustees of Near East Relief, 1944.

Nilsson, M. P. Griechische Feste von religiöser Bedeutung. Leipzig: Teubner, 1906.

Opler, M. E. "Family, Anxiety, and Religion in a Community of North India," in M. K. Opler, ed., *Culture and Mental Health.* New York: Macmillan, 1959.

Opler, M. K. "Dream Analysis in Ute Indian Therapy," in M. K. Opler, ed., *Culture and Mental Health.* New York: Macmillan, 1959.

Paul, B., ed. Health, Culture, and Community: Case Studies of Public Reactions to Health Programs. New York: Russell Sage Foundation, 1954.

Pausanius. Description of Greece. London: Heinemann, 1937. 5 vols.

Ploss, H. Das Kind in Brauch und Sitte der Völker. Leipzig: Grieben, 1884. Vols. I and II.

———. Das Weib in der Natur und Völkerkunde. Edited by M. Bartels. Leipzig: Grieben, 1905. Vols. I and II.

Politis, M. G. "Laografika Symmikta," Vol. C, Publication No. 6 of the Laographic File of the Academy of Athens, 1883 (in Greek).

Redfield, R. The Primitive World and its Transformations. Ithaca, N.Y.: Cornell Univ. Press, 1953.

Reider, N. "The demonology of modern psychiatry," *Am. J. Psychiat.*, 1955, **111**, 851–56.

Ridgeway, W. The Early Age of Greece. Cambridge, Eng.: Cambridge Univ. Press, 1931.

Rivers, W. H. R. Medicine, Magic, and Religion. New York: Harcourt, Brace, 1924.

Rodd, R. The Customs and Lore of Modern Greece. London: David Stott, 1892.

Roheim, G. "The Evil Eye," in S. Lorand, ed., *The Yearbook of Psychoanalysis*. New York: Int. Universities Press, 1953.

Rolleston, J. D. "Ophthalmic folklore," *Brit. J. Ophthalmol.*, 1942, **26**, 481–502.

Rosen, G. "What is social medicine?," *Bull. Hist. Med.*, 1947, **21**, 674–733.

Rosenfeld, L., B. Crowther, and M. C. Ring. Content and Technical Methods. Washington, D.C.: Div. Public Health Methods, U.S.P.H.S., 1952.

Ross, F., L. Fry, and E. Sibley. The Near East and American Philanthropy. New York: Columbia Univ. Press, 1929.

Rostovzeff, M. The Social and Economic History of the Hellenistic World. Oxford: Clarendon Press, 1941. Vols. I–III.

Roth, J. "Ritual and magic in the control of contagion," *Am. Soc. Rev.*, 1957, **22**, 310–14.

Sanders, I. Balkan Village. Lexington: Univ. of Kentucky Press, 1949.

Saunders, L. Cultural Difference and Medical Care. New York: Russell Sage Foundation, 1954.

Saunders, L., and G. Hewes. "Folk medicine and medical practice." *J. Med. Educ.*, 1953, **28**, 43–46.

Schmidt, B. Das Volksleben der Neugriechen und das Hellinische Altertum. Leipzig: Teubner, 1871.

———. Griechische Märchen, Sagen, und Volkslieder. Leipzig: Teubner, 1877.

Seligman, K. The Mirror of Magic. New York: Pantheon Books, 1948.

Siegel, B. "Medical practitioners and social structure in Brazil and Portugal," *Sociologia*, **20**, December 1958.

Sigerist, H. E. A History of Medicine. Vol. I: Primitive and Archaic Medicine. New York: Oxford Univ. Press, 1951.

Simmons, L. W., and H. G. Wolff. Social Science in Medicine. New York: Russell Sage Foundation, 1954.

Spicer, E. H. Human Problems in Technological Change. New York: Russell Sage Foundation, 1952.

Spiro, M. E., and R. G. D'Andrade. "A cross-cultural study of some supernatural beliefs," *Am. Anthropologist,* 1958, **60**, 456–66.

Stavrianos, L. S. The Balkans since 1453. Holt, 1948.

Stein, M. R. The Eclipse of Community: An Interpretation of American Studies, Princeton, N.J.: Princeton Univ. Press, 1960.

Stephan, K. Psychoanalysis and Medicine. Cambridge, Eng.: Cambridge Univ. Press, 1935.

Stephanides, C. S. "Where lack of fuel means food shortage," *Foreign Agriculture,* 1940, **14**, 95–98.

Stephens, W. N. "A cross-cultural study of menstrual taboos," *Gen. Psychol. Monogr.,* 1961, **64**, 385–416.

Swanson, G. Birth of the Gods: The Origin of Primitive Beliefs. Ann Arbor: Univ. of Michigan Press, 1960.

Thompson, Stith. "Motif-Index of folk literature," *FF Communications* (Helsinki: Suomalainen Tiedeakatemia), Nos. 106–109, 116–17, 1932–37.

Thorndike, Lynn. History of Magic and Experimental Science. New York: Macmillan, 1939.

Toynbee, A. J. Greek Civilization and Character. Boston: Beacon, 1950.

United Nations. Report of the FAO Mission in Greece. Washington, D.C.: FAO, 1947.

Ware, K. E., S. Fisher, and S. Cleveland. "Body image boundaries and adjustment to poliomyelitis," *J. Abnorm. Soc. Psychol.,* 1957, **55**, 88–93.

Westermarck, E. Ritual and Belief in Morocco. London: Macmillan, 1926.

Whipple, C. E. "The agriculture of Greece," *Foreign Agriculture,* 1944, **8**, 75–96.

Whiting, J. W. M. "Sorcery, Sin, and Superego: A Cross-Cultural Study of Some Mechanisms of Social Control," in M. R. Jones, ed., *Nebraska Symposium on Motivation.* Lincoln, Neb.: Univ. of Nebraska Press, 1959.

Whiting, J. W. M., and I. R. Child. Child Training and Personality: A Cross-Cultural Study. New Haven, Conn.: Yale Univ. Press, 1953.

Whitman, C. H. Homer and the Heroic Tradition. Cambridge, Mass.: Harvard Univ. Press, 1958.

Widening Horizons in Medical Education: A study of the teaching of social and environmental factors in medicine. Report to the Joint Comm. of the Assn. of Am. Med. Colleges and the Am. Assn. of Med. Social Workers. New York: Commonwealth Fund, 1948.

Zarlock, S. P. "Magical thinking and associated psychological reactions to blindness," *J. Consult. Psychol.,* 1961, **25**, 155–59.

Index

Index